SYLLABUS
OF
A COMPLETE COURSE
OF LECTURES
ON
CHEMISTRY

SYLLABUS
OF
A COMPLETE COURSE OF LECTURES
ON
CHEMISTRY

INCLUDING ITS APPLICATION TO

THE ARTS, AGRICULTURE, AND MINING.

PREPARED FOR THE USE OF THE GENTLEMEN CADETS AT

THE HONORABLE EAST INDIA COMPANY'S MILITARY SEMINARY, ADDISCOMBE

E. SOLLY

GYAN

Published by

Gyan Publishing House
5, Ansari Road
Daryaganj, New Delhi-110002
Phone: 011-47034999, 9811692060
E-mail: books@gyanbooks.com

Distribution Network
gyanbooks.com
India, USA, Canada, UK, Australia

ISBN: 978-93-6780-299-1 (HB)

First Published, 1849

2nd Impression 2024

Printed at: Gyan Press, Delhi.

SYLLABUS OF A COMPLETE COURSE OF LECTURES ON CHEMISTRY

Author: E.SOLLY

SYLLABUS

OF

A COMPLETE COURSE OF LECTURES

ON

CHEMISTRY,

INCLUDING ITS APPLICATION TO

THE ARTS, AGRICULTURE, AND MINING.

PREPARED FOR THE USE OF THE GENTLEMEN CADETS AT

THE HONORABLE EAST INDIA COMPANY'S MILITARY SEMINARY, ADDISCOMBE.

BY

PROFESSOR E. SOLLY,

OF THE HORTICULTURAL SOCIETY OF LONDON; FELLOW OF THE ROYAL, GEOLOGICAL, LINNÆAN, CHEMICAL, AND ANTIQUARIAN SOCIETIES OF LONDON; FELLOW OF THE SOCIETY FOR THE ENCOURAGEMENT OF ARTS AND MANUFACTURES; HONORARY MEMBER OF THE ROYAL AGRICULTURAL SOCIETY; HONORARY MEMBER OF THE POLYTECHNIC SOCIETY, CORNWALL; LECTURER ON CHEMISTRY IN THE H. E. I. C. MILITARY SEMINARY AT ADDISCOMBE, ETC. ETC.

REVISED, WITH ADDITIONS,

BY THE AUTHOR

OF "CHEMICAL AND PHARMACEUTICAL MANIPULATIONS," ETC. ETC.

LONDON:
LONGMAN, BROWN, GREEN AND LONGMANS.
PHILADELPHIA:
REPRINTED BY HENRY C. BAIRD,
SUCCESSOR TO E. L. CAREY,
S. E. CORNER FIFTH AND MARKET STREETS.
1850.

PREFACE.

The following Syllabus is intended to present to the Student a sketch of the Science of Chemistry, and its application to the Arts, Mining, and Agriculture. It includes a methodical arrangement of the laws of the science, as well as those collateral branches of knowledge which are essential to the study of Chemistry; together with systematic lists of all those substances which are of practical importance.

The arrangement, which I have adopted, differs slightly from those usually followed, inasmuch as organic chemistry, in place of forming a distinct head, is brought in after carbon; by so doing, a good deal of inconvenience and repetition is avoided; the various allied compounds being thus brought together, and classed under their proper heads, in place of being part amongst the organic, and part amongst the inorganic compounds.

The lists of metallic salts, etc., have been shortened as much as possible, that of the first metal being made tolerably copious, whilst in the succeeding ones only the more important compounds are mentioned.

For the convenience of reference, a short Index has been added; it contains only those substances less easily found in the systematic arrangement.

E. S.

February, 1849.

EDITOR'S PREFACE.

The present work is designed to occupy a vacant place in the libraries of Chemical text-books. It is admirably adapted to the wants of both teacher and pupil; and will be found especially convenient to the latter, either as a companion in the class room or as a remembrancer in the study. It gives, at a glance, under appropriate headings, a classified view of the whole science, which is at the same time compendious and minutely accurate: and its wide margins afford sufficient blank space for such manuscript notes as the student may wish to add during lectures or recitations.

The almost indispensable advantages of such an impressive aid to memory are evident to every student, who has used one, in other branches of study. Therefore, as there is now no Chemical Syllabus, we have been induced by the excellences of this work to recommend its republication in this country; confident that an examination of the contents, will produce full conviction of its intrinsic worth and usefulness.

GENERAL ARRANGEMENT.

PART I.—PHYSICS.

PART II.—CHEMISTRY OF THE NON-METALLIC ELEMENTS.

PART III.—CHEMISTRY OF THE METALLIC ELEMENTS.

PART I.

PHYSICS OF CHEMISTRY.—OF MATTER AND FORCE.

SECT. I.—OF MATTER.

1. THEORETICAL CONSTITUTION OF MATTER.
2. PROPERTIES.
   - A. Impenetrability.
   - B. Indestructibility.
   - C. Inertia.
   - D. Divisibility.

SECT. II.—OF FORCE.

SUBSECT. I.—ATTRACTION.

1. ATTRACTION OF AGGREGATION.
   - A. Solids.
     - i. Cohesion.
       - *a.* Hardness.
       - *b.* Tenacity.
       - *c.* Flexibility.
     - ii. Crystallization.
       - *a.* Arrangement of forms.
       - *b.* Modes of obtaining crystals.
       - *c.* Method of measuring crystals. Goniometers.
   - B. Fluids and gases.
   - C. Elasticity in solids, fluids, and gases.
     - i. Vibration.
       - *a.* Cause of sound—velocity in different media.
       - *b.* Molecular changes produced by.
   - D. Dimorphism.—Existing in two distinct forms.

2

2. Attraction of Gravitation.
   - A. Absolute weight.
     - i. The balance—double weighing.
   - B. Relative weight. Specific gravity.
     - i. Of solids and powders. Hydrostatic balance.
     - ii. Of liquids. Hydrometers. Specific gravity bottles.
     - iii. Of gases. Precautions as to heat and pressure.
     - iv. Of vapors.

3. Attraction of Heterogeneous Substances.
   - A. Between solids.
     - i. Friction.
     - ii. Cements. Glue.
   - B. Between solids and liquids.
     - i. Wetting.
     - ii. Capillary action.
     - iii. Solution. Point of saturation.
   - C. Between liquids.
     - i. Endosmosis and exosmosis.
     - ii. Solution and penetration.
   - D. Between solids and gases, or solids and vapors.
     - i. Ebullition and effervescence.
   - E. Between liquids and gases.
     - i. Frothing.
   - F. Between gases.
     - i. Diffusion, effusion, and transpiration.

4. Attraction of Affinity. Chemical Affinity.
   - A. Effects of chemical affinity.
     - i. Changes in properties invariably produced.
   - B. Combination, or the union of substances.
     - i. Conditions requisite.—Affinity, contact, temperature.
     - ii. Interfering causes.
       - *a.* Attraction of aggregation.
   - C. Definite proportions, invariable combining weights.
     - i. Atomic theory.
     - ii. Relation of volume and weight.
       - *a.* System of nomenclature.
       - *b.* Chemical symbols.

iii. Isomerism. Similar compounds having different properties.
iv. Isomorphism. Similarity of atomic composition and crystalline form.

D. Decomposition, or separation into simpler substances.
  i. Conditions requisite.
    a. Interfering causes, aggregation, &c.
  ii. Spontaneous decomposition.
    a. Influence of attraction, heat, light, electricity.
  iii. Decomposition by Superior Affinity.
    a. Double decomposition.
  iv. Decomposition by Example—Fermentation.
  v. Decomposition by Presence—Catalytic Action.
  vi. Influence of Heterogeneous Attraction.
    a. Nascent gases, &c.

Subsect. II.—Heat.

1. General Nature and Effects of Heat. Temperature.

A. Sources of heat.
  i. Solar. Its difference from terrestrial.
  ii. Mechanical. Friction. Compression.
  iii. Chemical. In combination. In decomposition.
    a. Organic. In animals. In plants.
  iv. Electrical.
  v. Molecular changes.

B. Expansion produced by heat.
  i. Solids. Nearly all expand when heated.
  ii. Liquids. Ditto; even more than solids.
  iii. Gases and vapors, the most expansible.
  iv. Exceptions to general rule. Water, ice, fusible metals, &c.

C. Measures of heat.
  i. Thermometers. Simple and differential. Registering. Mercurial and spirit.
  ii. Pyrometers. Wedgwood's. Daniell's. Prinsep's.

D. Conduction of heat.
  i. Passage of heat through solids.
    a. Decrepitation. Caused by unequal expansion.

ii. Passage of heat through liquids.
iii. Passage of heat through gases.
a. Convection, or carrying of heat.
b. Natural equalization of temperature.
c. Ventilation.

2. Latent Heat.
A. Fusion, or Thawing. Heat absorbed.
B. Boiling, or Vaporization.
i. Spheroidal state.
ii. Cold of evaporation. Heat absorbed.
C. Condensation and Solidification. Heat set free.
i. Sublimation.
ii. Distillation. Stills. Alembics. Retorts. Influence of air; distillation in vacuo.
iii. Theory of the steam-engine.

3. Specific Heat.
A. In solids. Different quantities to produce same effect.
B. In liquids.
C. In gases.
D. Relation to volume.

4. Radiant Heat.
A. Solar heat.
i. Transmission.
ii. Reflexion. Concave mirrors.
iii. Refraction. Burning lenses.
iv. Absorption.
B. Terrestrial heat.
i. Transmission. Diathermanous substances.
ii. Reflexion.
iii. Refraction.
iv. Absorption. Reception and emission.
v. Radiation. Influence of surface.

Subsect. III.—Light.

1. Theories of Light.
A. Emission, or Newtonian theory.
B. Undulation, or Huygenian theory.

2. **Sources of Light.**
   - A. The sun.
   - B. Heat. Ignition. Incandescence.
   - C. Electricity.
   - D. Chemical action. Combination, and decomposition.
   - E. Phosphorescence.
     - i. Solar phosphori.
     - ii. Phosphorescence caused by heat.
     - iii. Organic phosphori.
   - F. Changes in molecular constitution.
     - i. Changes of dimorphous bodies. Crystallization.

3. **Transmission of Light in Space. Its great Velocity.**

4. **Reflexion.**
   - A. From plane surfaces. Mirrors.
   - B. From curved surfaces.

5. **Refraction.**
   - A. Modified by density of media. Lenses.
   - B. Modified by chemical nature.
     - i. Mode of ascertaining refractive power of substances.

6. **Double Refraction and Polarization.**
   - A. By refraction.
   - B. By reflexion.
   - C. Circular polarization. Analysis of sugars by.
     - i. Action of different substances on polarized rays.

7. **Decomposition of Light.**
   - A. By refraction. The prism. Frauenhofer's lines.
   - B. By absorption.
     - i. Cause of colors. Monochromatic light.

8. **Chemical Powers of Light.**
   - A. Chemical rays.
     - i. Principles of photography.

9. **Measures of Light. Photometry.**
   - A. By comparison of shadows. Rumford's Photometer.
   - C. Wheatstone's Photometer.
   - C. Photographic Photometers.

10. **Relations of Heat and Light to each other.**

Subsect. IV.—Electricity.

1. Electricity of Tension. Frictional Electricity.
   - A. Attraction.
   - B. Repulsion.
   - C. Electrical theories. Du Fay's and Franklin's.
   - D. Conduction of electricity.
     - i. Conductors, or non-electrics. Velocity of conduction.
     - ii. Non-conductors, or electrics. Insulation.
   - E. Sources of electricity of tension.
     - i. Electrical machines. Construction and use.
     - ii. Atmospheric electricity.
       - *a.* Thunder storm.
       - *b.* Aurora borealis. Fire of St. Elmo.
     - iii. Electricity of steam.
     - iv. Electricity of cleavage.
   - F. Induction. Quantity and Intensity. Dielectrics.
     - i. Leyden jar. Electrical batteries.
     - ii. Electrophorus.
     - iii. Condenser.
   - G. Distribution.
     - i. Surface—accumulation in particular forms.
     - ii. Points—their power of receiving and dispersing. Flame.
   - H. Effects.
     - i. Light. Spark. Brush. Light in vacuo.
     - ii. Heat. Inflammation. Ignition.
     - iii. Mechanical disruption. Destruction of buildings.
       - *a.* Lightning conductors. Application to powder magazines, houses, ships, &c.
     - iv. Physiological effects. Shock. Medical electricity.
     - v. Magnetism produced.
     - vi. Chemical actions caused.—Ozone.
   - I. Measures of electricity.
     - i. Electrometers. Bennet's. Coulomb's. Harris's, &c.
     - ii. Electroscopes.

2. Electricity of Quantity. Current, or Chemical Electricity.

- A. Sources.
  - i. Simple circuit.
  - ii. Voltaic or galvanic batteries. Constant batteries.
  - iii. Difference of intensity and quantity. Dry piles.
- B. Theories.
  - i. Chemical hypothesis.
  - ii. Contact theory.
    - *a.* Mathematical expression. Ohm's formula.
- C. Conduction.
  - i. Conductors.
  - ii. Non-conductors.
  - iii. Resistance. The Rheostat.
- D. Effects of current electricity.
  - i. Heat. Ignition.
    - *a.* Application to firing mines.
    - *b.* Application to submarine blasting, as practiced by General Pasley.
  - ii. Light. Deflagration of metals.
  - iii. Physiological effects. Shock.
  - iv. Magnetism produced.
    - *a.* Electro-magnets.
    - *b.* Helices. Coils. Accumulation.
    - *c.* Electro-magnetic rotations.
    - *d.* Electric telegraphs, clocks, alarums.
  - v. Secondary currents.
  - vi. Chemical effects produced.
    - *a.* Electrolysis. Definite action of electricity.
    - *b.* Anions and cations. Determination of ions.
      - *aa.* Purification of water, &c.
    - *c.* Principles of electro-metallurgy.
      - *aa.* Secondary decompositions.
- E. Measures of current electricity.
  - i. Magnetic. Galvanometers.
  - ii. Chemical. Voltameters.

3. Magneto-electricity.

- A. Sources. Magneto-electric machines.
- B. Theory of its evolution.

C. Effects. Luminous, heating, physiological and chemical effects.

4. Thermo-electricity.
 A. Sources. Thermo-electric piles and batteries.
 B. Effects. Luminous, heating, physiological and chemical effects.
  i. Thermo-electric thermoscopes.
  ii. Production of heat and cold in compound metallic arrangements, by electric currents.

5. Animal Electricity.
 A. Sources. Gymnotus, Torpedo, and other animals.
 B. Effects produced. Heat, light, magnetism, physiological and chemical actions.
 C. Connection with nervous influence, and muscular action.

Subsect. V.—Magnetism.

1. Magnetism and Diamagnetism.
 A. Magnetism of iron, nickel, cobalt, platinum, palladium, titanium, and many of their salts.
 B. Diamagnetism of bismuth, antimony, zinc, tin, cadmium, mercury, silver, copper, gases, &c.

2. Attraction and Repulsion.

3. Terrestrial Magnetism.
 A. The Compass. Dip. Variation.
 B. Magnetic disturbances, and storms.
 C. Natural magnets, or loadstones.

4. Magnetic Induction.

5. Electro-Magnetism.
 A. Modes of making artificial magnets.

6. Effects of Heat on Magnetism.

7. Influence of Magnetism on Light.
 A. Rotation of a ray of light in the axis of a helix, conveying a current of electricity.

8. General Relations of Magnetism, Heat, Light, Electricity, and Chemical Action with each other, and with Motion.

PART II.

CHEMISTRY OF THE NON-METALLIC ELEMENTS.

Sect. I.—Analysis.

1. Principles of Analysis.
   - A. Definition of an Element, or simple substance.
   - B. Binary compounds.
   - C. Ternary and quaternary compounds.
2. Modes of Testing.
   - A. Apparatus required.
   - B. Use of balance; precautions.
   - C. Use of heat.
   - D. Application of solvents.
   - E. Employment of tests and precipitants.
3. Qualitative Examination.
4. Quantitative Analysis.
5. Calculation of Atomic Constitution.
6. Arrangement of Elements into Groups.

7. List of the Elements, or Undecomposed Substances.

Name.	Symbol.	Equivalent.	Name.	Symbol.	Equivalent.
1. Aluminum	Al	14	32. Molybdenum	Mo	48
2. Antimony	Sb	129	33. Nickel	Ni	28
3. Arsenic	As	75	34. Niobium	Nb	—
4. Barium	Ba	69	35. Nitrogen	N	14
5. Bismuth	Bi	213	36. Osmium	Os	100
6. Boron	B	11	37. Oxygen	O	8
7. Bromine	Br	78	38. Palladium	Pd	54
8. Cadmium	Cd	56	39. Pelopium	Pl	—
9. Calcium	Ca	20	40. Phosphorus	P	32
10. Carbon	C	6	41. Platinum	Pt	99
11. Cerium	Ce	46	42. Potassium	K	40
12. Chlorine	Cl	36	43. Rhodium	R	52
13. Chromium	Cr	28	44. Ruthenium	Rt	—
14. Cobalt	Co	30	45. Selenium	Se	40
15. Columbium	Ta	185	46. Silicon	Si	21
16. Copper	Cu	32	47. Silver	Ag	108
17. Didymium	D	—	48. Sodium	Na	23
18. Erbium	E	—	49. Strontium	Sr	44
19. Fluorine	Fl	19	50. Sulphur	S	16
20. Glucinum	G	7	51. Tellurium	Te	64
21. Gold	Au	200	52. Terbium	Tb	—
22. Hydrogen	H	1	53. Thorium	Th	60
23. Iodine	I	126	54. Tin	Sn	59
24. Iridium	Ir	99	55. Titanium	Ti	24
25. Iron	Fe	28	56. Tungsten	W	100
26. Lantanum	La	44	57. Uranium	U	60
27. Lead	Pb	104	58. Vanadium	V	68
28. Lithium	Li	7	59. Yttrium	Y	32
29. Magnesium	Mg	12	60. Zinc	Zn	32
30. Manganese	Mn	28	61. Zirconium	Zr	33
31. Mercury	Hg	100			

Sect. II.—Oxygen, Chlorine, Iodine, Bromine, Fluorine.

1. Oxygen. Symb. O. Eq. 8.

A. Natural sources of.—Air, water, all rocks, &c. Discovered by Priestley. Derivation of name.

B. Management of gases.—Pneumatic trough and apparatus.

C. Modes of obtaining oxygen.

i. By heating chlorate of potash; oxide of manganese; red oxide of mercury; nitre; a mixture of chloride of lime and oxide of manganese, &c. Theory of each process. Collected over water. Gasometers.

D. Properties of oxygen.

i. Physical characters.—Permanently elastic, transparent, colorless gas; sp. gr. 1.109.

ii. Chemical characters.—Powerful supporter of combustion and respiration; forms oxides and acids; has no action on vegetable colors; tasteless, inodorous. Solubility.

2. CHLORINE. Symb. Cl. Eq. 36. (35.50.) Discovered by Scheele. Derivation of name.

A. Natural sources.—Common salt, &c.

B. Processes for preparing it.—Heating hydrochloric acid with peroxide of manganese; heating common salt, oxide of manganese, and sulphuric acid. Collected over tepid water. Theory of each process.

C. Properties, physical and chemical.—Elastic, transparent, pale yellow gas; condensed under 4.5 atmospheres to oily fluid; soluble in cold water; suffocating, poisonous gas; sp. gr. 2.470; ditto, liquid chlorine, 1.33; strong affinity for metals; forms chlorides; when dry, has little action on vegetable colors; when moisture is present, bleaches them. Solubility.

D. Uses in the arts.—Bleaching, and disinfecting.

E. Compounds of chlorine with oxygen.

i. Protoxide, hypochlorous acid, $Cl\,O$. An orange yellow liquid which boils at 68°; obtained by the action of dry chlorine on peroxide of mercury. Its vapor is yellow; sp. gr. 2.977; is readily absorbed by water, and is decomposed with great facility by heat. Its aqueous solution is likewise very prone to decompose.

ii. Chlorous acid, $Cl\,O_3$. A greenish-yellow gas, not liquefied by a cold of 5°; soluble in water; decomposed by light, and also by a temperature of 140°. It may be obtained by heating a mixture of chlorate of potash, arsenious acid, and nitric acid.

iii. Peroxide, hypochloric acid, $Cl\,O_4$. A red liquid which boils at 68°, forming a deep yellow gas; obtained by the action of sulphuric acid on chlorate of potash. The gas is soluble in water, and decomposed by mercury; sp. gr. about 2.337; slowly decomposed by light; explodes when heated above 200°.

iv. Chloric acid, $Cl\ O_5$. Formed by the action of chlorine on a base, as potash; does not exist independent of a base, except in solution, which is a thick, syrup-like, sour liquid; decomposed by a heat above 100°.

v. Perchloric acid, $Cl\ O_7$. Formed by the action of sulphuric acid on the chlorates. It has not been obtained quite anhydrous. Its hydrate is a white, volatile, crystalline solid, deliquescent, and very soluble in water, forming a colorless, fuming, sour liquid, sp. gr. 1.65, which boils about 400°.

3. IODINE. Symb. I. Eq. 126. (126.36.) Discovered by Courtois. Derivation of name.

A. Native sources.—Sea water, rough sea salt, various marine plants and animals, some mineral springs, certain rare lead and silver ores.

B. Processes for obtaining it.—Heating the uncrystallizable residue of kelp lixivium with oxide of manganese and sulphuric acid. Theory of the operation.

C. Physical and chemical properties.—Bluish black, semi-metallic-looking, friable, crystalline solid; easily volatile; fuses at 225°; boils at 350°; forms violet vapor, blues starch; combines with metals, &c., to form iodides; sp. gr. 4.948:—of its vapor 8.716. Solubility.

D. Uses in the arts.—Dyeing, and in medicine.

E. Compounds with oxygen.

i. Hypoiodic acid, $I\ O_4$.

ii. Iodic acid, $I\ O_5$. A white solid, obtained by the action of strong nitric acid on iodine. It is soluble in ether, alcohol, and water, forming a strong sour solution. It crystallizes from its solution, as a hydrate, $I\ O_5 + H\ O$. Iodic acid fuses at about 500°, and is decomposed by a higher temperature.

iii. Periodic acid, $I\ O_7$. A white, solid hydrate; is obtained by decomposing periodate of silver by water. It forms white crystals, which fuse at 266°, and are decomposed by a higher temperature. They readily dissolve in water.

F. Compounds with chlorine.

i. Protochloride, I Cl. An orange, oily liquid, obtained by distilling a mixture of iodine and chlorate of potash. It is soluble in water.

ii. Terchloride, $I\ Cl_3$. A yellow, fusible, crystallizable solid, obtained direct.

4. Bromine. Symb. Br. Eq. 78. (78.26.) Discovered by Balard. Origin of name.

A. Native sources.—Sea water, mineral springs, the mother waters of kelp, native bromide of silver, &c.

B. Process for obtaining it.—Action of chlorine on bittern, and agitation with ether, which dissolves the bromine. Theory of the process.

C. Physical and chemical properties.—Deep red, volatile liquid; solidifies at—8°; boils at 116°.5; sp. gr. 2.966; vapor red, and of sp. gr. 5.39; forms bromides. Solubility.

D. Compounds with oxygen and chlorine.

i. Bromic acid, $Br\ O_5$. Exists only as an aqueous solution, obtained like chloric acid.

ii. Chloride of bromine, $Br\ Cl_5$. A red, yellow liquid, obtained direct. It is soluble in water, but decomposed by alkaline solutions.

5. Fluorine. Symb. Fl. Eq. 19. (18.70.) Discoverer. Origin of name.

A. Native combinations.—Fluor spar, &c.

B. Not isolated, only known in combination.

Sect. III.—Hydrogen, Nitrogen, Sulphur, Phosphorus, Selenium, Boron, Silicon and Carbon.

1. Hydrogen. Symb. H. Eq. I. Discovered by Cavendish. Origin of name.
   - A. Native sources.—Water, &c.
   - B. Processes for obtaining it.—Decomposition of steam by red-hot iron; ditto of water, by potassium, or by zinc and sulphuric acid; self-regulating reservoir collected over water. Theory.
   - C. Physical and chemical properties.—Colorless, transparent, permanently elastic gas; extinguishes flame, but burns itself; very light gas; sp. gr. to air, 0.069; 100 cu. in. weigh 2·14 grains; strong affinity for oxygen; powerful deoxidizer; when pure, has neither taste, smell, nor action on vegetable colors.
   - D. Uses in the arts.—Hydro-oxygen blowpipe; Drummond's light; balloons.
   - E. Compounds with oxygen.
     - i. Water, oxide of hydrogen, H O.
       - *a.* Composition and decomposition of water.
       - *b.* Physical and chemical properties of water.—Freezes at 32°; boils at 212°; combines to form hydrates.
       - *c.* Impurities, and purification of water.—Modes of ascertaining the purity of water; hard and soft waters; springs and wells; natural varieties of water. Artificial waters.
     - ii. Peroxide of hydrogen, oxywater, $H\,O_2$. Colorless, inodorous liquid; sp. gr. 1.45; instantly decomposed by the metals, and spontaneously at common temperatures.
   - F. Compound with chlorine.—Hydrochloric or muriatic acid, Cl H.
     - i. Process for obtaining it. — Action of sulphuric acid on common salt.
     - ii. Physical and chemical properties.—Transparent, colorless gas; fumes in the air, absorbing

moisture; collected over mercury; extinguishes flame; sp. gr. 1.262; powerful acid; reddens litmus and also turmeric paper; may be liquefied by a pressure of 40 atmospheres; very soluble in water; neutralizes bases, forming chlorides or muriates; used in the arts, medicine, &c.

G. Compound with iodine.—Hydriodic acid, I H.

i. Processes for obtaining.—Action of water on iodide of phosphorus; action of iodine on sulphuretted hydrogen.

ii. Properties.—Transparent, colorless gas; fumes in the air; soluble in water, and decomposed by mercury; strongly acid; extinguishes flame; condensible under pressure, and solidifiable by cold; sp. gr. of gas, 4.4024.

H. Compound with bromine.—Hydrobromic acid, Br H.

i. Process for obtaining.—Same as hydriodic acid.

ii. Properties.—Colorless, transparent gas; very soluble in water; strongly acid; may be collected over mercury.

I. Compound with fluorine.—Hydrofluoric acid, Fl H.

i. Process for obtaining.—Distillation of sulphuric acid and fluor spar (fluoride of calcium), and subsequent decomposition of the hydrated acid by dry phosphoric acid.

ii. Properties.—Colorless, transparent gas; fumes in air; highly corrosive acid; very soluble in water; acts strongly on all compounds containing silica.

iii. Used in etching glass, &c.

2. NITROGEN, or AZOTE. Symb. N. Eq. 14. Discoverer. Origin of name.

A. Native sources.—The atmosphere, animal matters, &c.

B. Processes for obtaining it.

i. Combustion of phosphorus in air, and subsequent removal of the phosphoric acid formed by washing; collected over water.

ii. Properties.—Colorless, transparent, permanently elastic gas; no taste or smell; no action on colored tests; extinguishes flame; a perfect non-supporter of combustion; exhibits no tendency to enter into combination with other substances; sp. gr. 0.971.

C. Mixture of nitrogen with oxygen.—Atmospheric air.

i. Physical properties of the air.—Taken as standard of density, whole weight equal to 14 lbs. per square inch. 100 cu. in. weigh 31.0117 grains. Air-pump. Barometer, its variations. Mountain Barometer.

ii. Chemical composition of the atmosphere.—By weight, 23 parts oxygen to 77 of nitrogen; by measure, 21 volumes of oxygen to 79 of nitrogen; about one-thousandth part of carbonic acid gas, and a variable proportion of moisture. Effects of moisture in the air. Hygrometers. Hygroscopes Daniell's, the wet bulb. Phenomena of dew, rain, and hail. Eudiometry. Analysis of air. Extraneous matter in the air, Miasm, Infection.

D. Compounds of nitrogen and oxygen.

i. Nitrous oxide (Protoxide, Laughing gas), $N\,O$. Colorless, transparent gas; obtained by heating nitrate of ammonia; collected over water; supports flame; produces remarkable effects when inhaled; sp. gr. 1.525; no action on test papers. Liquefies at 0° under a pressure of about 30 atmospheres.

ii. Nitric oxide (Deutoxide), $N\,O_2$. Colorless, transparent gas; obtained by acting on dilute nitric acid by copper; collected over water; extinguishes some flames; seizes oxygen in contact with air, and forms red fumes of nitrous acid gas; sp. gr. 1.041; has no action on colored tests.

iii. Hyponitrous acid, $N\,O_3$. Obtained with difficulty, by mixing 45 parts of water, 92 of nitrous acid, and carefully distilling; an orange vapor; readily soluble in water; combines with bases, to form hyponitrites; reddens vegetable blue tests.

iv. Nitrous acid, hyponitric acid, $N\ O_4$. Obtained by mixing 1 volume of oxygen with 2 volumes of dry nitric oxide in an exhausted receiver; a deep red vapor; condensible to an orange liquid at 0°, and solidifiable by a lower temperature; the liquid boils at 82°, and has a sp. gr. 1.45; that of the vapor is 3.19; an acid, but does not form salts with bases, being decomposed and hyponitrites and nitrates formed; reddens blue tests.

v. Nitric acid, $N\ O_5$. Does not exist pure in the separate state, but always with a base, or with water as a hydrate; its aqueous solution, Aqua fortis, obtained by distilling nitre (nitrate of potash) with sulphuric acid, $N\ O_5 + H\ O$; a colorless, intensely acid liquid; sp. gr. about 1.45; a powerful oxidizer; combines with bases to form nitrates; reddens blues strongly; exists in nature as nitrate of potash, etc.; formed in the air by electric sparks, flashes of lightning; used in the arts to oxidize and dissolve metals; of great importance in dyeing, metallurgy, and medicine. Fuming acid.

E. Compound of nitrogen and chlorine.—Chloride of nitrogen, oil of azote, $N\ Cl_3$. Obtained by acting on a solution of sal-ammoniac by chlorine; a yellow, oily-looking fluid, insoluble in water; requires great care in its preparation; violently explosive; distils at 160°; explodes at 212°; sp. gr. 1.65.

F. Compound of nitrogen and iodine.—Iodide of nitrogen, $N\ I_3$. Obtained by acting on an aqueous solution of ammonia by iodine; a black solid substance; extremely explosive, detonating on the slightest touch.

G. Compound with bromine.—Bromide of nitrogen, $N\ Br_3$. Obtained by acting on chloride of nitrogen by bromide of potassium under water; a dark red heavy volatile liquid, exceedingly explosive.

H. Compound with hydrogen.—Ammonia, $N\ H_3$.

i. Exists in nature, in the air, rain water, &c.; formed

by the decomposition of all organic substances containing nitrogen; destructive distillation of animal matter; obtained pure by heating muriate of ammonia with quick lime.—Physical and chemical properties. A transparent, colorless gas, condensible to a liquid by a pressure of 6.5 atmospheres, solidifiable by intense cold. The gas has a powerfully pungent, peculiar odor, and is strongly alkaline, hence called the volatile alkali; reddens vegetable yellows, and greens blues. It is absorbed by water, which dissolves about 700 times its volume (*aqua ammoniæ*); may be collected over mercury; extinguishes flame; has a density of 0.589; readily combines with acid to form salts.

ii. Hypothetical composition of Ammonia compounds.

a. Amidogen, or Amogen, $N H_2$. It has never been obtained separate, only in combination with metals, as amides, and in organic compounds.

b. Ammonium, $N H_4$. Likewise never isolated. Supposed by some to exist in the ammonia compounds in its independent state, or as an oxide $N H_4 O$.—Remarkable action of amalgam of potassium on moist sal ammoniac.

iii. Ammonia and hydrochloric acid.—Sal-ammoniac, $N H_3 Cl H$. Preparation, and properties.—Exists native near volcanoes; obtained by heating a mixture of crude sulphate of ammonia with common salt; may be obtained direct and pure by mixing ammonia and muriatic acid gas; a white fibrous or crystalline semi-transparent solid; volatile at a temperature above 212°; sp. gr. about 1.48; easily soluble in water. Manufacture—from salt of hartshorn, combustion of dung, from gas liquor.—Uses in the arts, &c., by workers in metal, preparation of dyers' spirits, solution of tin, &c.; as manure.

iv. Ammonia and nitric acid.—Nitrate of ammonia, $NH_3 NO_5 + HO$. Obtained direct; a white, crystalline salt, slightly deliquescent; easily soluble in water; fusible at 230°; boils at 350°; resolved into water and nitrous oxide at 400°; sp. gr. 1.7; contains 1 eq. of water of hydration. Its decomposition by heat thus represented:— $NH_3 NO_5 + HO = 2 NO$ and $4 HO$.

v. Ammonia and sulphur.—Four compounds.

3. Sulphur, or Brimstone. Symb. S. Eq. 16. Discoverer. Origin of name.

A. Exists in nature nearly pure in Sicily, India, China, South America; generally in volcanic districts; combined with metals or sulphurets, in most parts of the world; also, as sulphuric acid, in various sulphates, like gypsum. Obtained pure by sublimation; obtained from the metallic sulphurets by heat, which vaporizes the sulphur.

B. Physical and chemical properties.—A pale, yellow, brittle, tasteless, inodorous solid; fusing at 232°; boiling at 750°; and rising in vapor at most ordinary temperatures; sp. gr. about 2.0. When deposited from a solution, or slowly condensed, it crystallizes; very combustible, burns at a very low temperature; sublimes into *flowers;* quite insoluble in water; used in the manufacture of gunpowder, matches, certain metallic sulphurets, &c.

C. Compounds with oxygen. Thionic series.

i. Hyposulphurous acid, $S_2 O_2$. Formed when sulphur is digested in an aqueous solution of a sulphite; has not been obtained separate; combines with bases to form hyposulphites.

ii. Sulphuretted hyposulphuric acid, (Sulphypo-sulphuric or trithionic acid,) $S_3 O_5$. Formed when sulphur is digested in solution of certain bisulphites; very little examined.

iii. Bisulphuretted hyposulphuric acid, (Bisulpho-hyposulphuric or tetrathionic acid,) $S_4 O_5$. By acting on hyposulphite of baryta by iodine, a bisulphuretted hyposulphite of baryta is obtained, which can be decomposed by sulphuric acid, and a solution of the acid itself obtained, as a sour liquid, which combines with bases, &c.

iv. Sulphurous acid, $S\,O_2$. Formed by burning sulphur in air or oxygen, or by heating sulphuric acid with mercury; a transparent gas, condensible under a pressure of two atmospheres, and solidifiable by cold. The gas extinguishes flame, is irrespirable, has a suffocating odor, is soluble in water, and may be collected over mercury. Its sp. gr. is 2.2105. Combines with bases to form sulphites. It possesses considerable bleaching powers, and is accordingly used in the arts, as a means of destroying colors.

v. Hyposulphuric acid (Dithionic), $S_2\,O_5$. Obtained by acting on deutoxide of manganese in water by sulphurous acid, when sulphate and hyposulphate of manganese are formed, which being decomposed by baryta, and the salt obtained, in turn decomposed by sulphuric acid, gives a solution of hyposulphuric acid. It has not been obtained anhydrous; it readily unites with bases to form hyposulphates.

vi. Sulphuric acid, $S\,O_3$. Formed by burning a mixture of sulphur and nitre, and condensing the fumes in water; obtained pure by distilling dry sulphate of iron. Nordhausen acid, $2\,(S\,O_3), H\,O$. Monohydrated acid of commerce, $S\,O_3, H\,O$.

Other hydrates $\begin{cases} S\,O_3,\ 2\,H\,O. \\ S\,O_3,\ 3\,H\,O. \end{cases}$

a. Physical and chemical properties.—A white, silky-looking, fibrous solid, having a very powerful affinity for water, fusible at about 75°, readily volatile, and having a density of about 1.97. Dissolved in water, it forms ordinary oil of vitriol, or sulphuric acid, which is a hydrate of the pure acid, and therefore represented by $S\,O_3 + H\,O$. It is a heavy, oily-looking liquid; sp. gr. 1.84; solidifying at — 15°, and boiling at 620°. It is an intensely acrid, powerful acid, destroying organic matter rapidly, combining with bases readily to form sulphates, and displacing almost all other acids, in consequence of the strong affinity which it has for bases.

b. Manufacture on the large scale.—By combustion of nitre and sulphur, formation of a compound of nitrous and sulphurous acids with water ($NO_4 + 2SO_2 + 2HO$), which by solution in water is resolved into $2SO_3$ and NO_2; concentration and distillation.

c. Uses in the arts.—Manufacture of soda, nitric acid, muriatic acid, chlorine, bleaching powder, metallurgy, sulphates, alum, &c.

D. Sulphuric acid and ammonia.—Sulphate of ammonia, $NH_3SO_3 + HO$. Formed direct; a white, soluble, crystallizable salt, manufactured by adding sulphuric acid to the ammoniacal liquor of the gas works, and purifying by crystallization; used as a source of ammonia compounds, and as a manure.

E. Sulphur and chlorine.—Chlorides of sulphur.

i. Subchloride, or dichloride of sulphur, S_2Cl. Formed by exposing excess of sulphur to chlorine; a yellow, brown liquid; sp. gr. 1.68, which boils at 280°; decomposed by water, muriatic acid, sulphur, and sulphurous acid being evolved.

ii. Chloride of sulphur, SCl. Formed when sulphur is heated in excess of chlorine; a greenish, orange liquid; fuming in the air; decomposed by water; boiling at 146°, and having a density of 1.60.

F. Sulphur and iodine, bromine &c. (analogous to chloride).

G. Sulphur and hydrogen.

i. Sulphuretted hydrogen.—Sulphydric or hydrothionic acid, SH.

a. Spontaneously generated by the decomposition of organic substances containing sulphur; evolved during putrefaction; obtained by acting on protosulphuret of iron by water and sulphuric acid.

b. Properties.—A colorless, transparent gas, condensible under a pressure of 17 atmospheres, and solidifiable; may be collected over water, which, however, gradually absorbs it; has a nauseous and disgusting odor; highly poisonous; sp. gr. 1.1786; extinguishes flame, but burns itself with a pale flame, forming water and sulphurous acid, sulphur being deposited; has little or no action on test-papers, but possesses feeble acid powers; instantly decomposed by chlorine, &c.; blackens salts of lead; combines with bases to form hydrosulphurets; that of ammonia is a useful test for metals.

ii. Bisulphuretted hydrogen, $H\ S_2$. Formed by pouring a strong solution of persulphuret of potassium into diluted muriatic acid; a heavy, yellow, oily-looking, viscid liquid; prone to decompose; sp. gr. 1.7.

H. Sulphur and nitrogen, $N\ S_3$. Formation—dry ammonia, acted on by perchloride of sulphur, yields an ammonia chloride of sulphur, which, decomposed by water, leaves a greenish yellow solid sulphuret of nitrogen.

4. Phosphorus. Symb. P. Eq. 32. Discovered by Brandt. Origin of name.

A. Exists as phosphoric acid, in apatite, in various rocks, in fertile soils, in most vegetable and animal substances. Obtained from bone earth, or from the native phosphate of lime, these are decomposed by sulphuric acid, and the resulting glass of phosphorus or phosphoric acid deoxidized by distillation with charcoal.—Precautions necessary to prevent its inflammation.

B. Physical and chemical properties.—It is a wax-like, nearly transparent, brittle yet flexible solid, fusing at

110°, and boiling at 572°; sp. gr. about 2.0; very combustible; exhales fumes, which are luminous in the dark; insoluble in water, soluble in ether, oils, &c.; is deposited in crystals from its solutions; unites with the metals, &c., to form phosphurets. Used chiefly in the manufacture of matches.

C. Compounds of phosphorus with oxygen.

i. Oxide of phosphorus, P_2O. A red solid substance, formed when phosphorus is imperfectly burnt; insoluble in water; burns at about 600°.

ii. Hypophosphorous acid, PO. Formed when baryta is boiled in water with phosphorus; the acid thus formed is combined with oxide of lead, and the lead salt decomposed by sulphuretted hydrogen; a thick, sour liquid results; it has not been obtained anhydrous; it combines with bases to form hypophosphites.

iii. Phosphorous acid, PO_3. Formed when phosphorus is burnt in rarefied air, or by acting on terchloride of phosphorus by water. The acid crystallizes with 3 equivalents of water; it unites with bases to form phosphites.

iv. Phosphoric acid, PO_5.

a. Formed when phosphorus is burnt in oxygen, or in free air; obtained by decomposing phosphate of lime by sulphuric acid.

b. Properties.—An inodorous, uncrystallizable, white solid, fusible at a red heat, and then transparent. It is deliquescent, very soluble in water, strongly acid, reddening vegetable blues, and unites to bases to form phosphates.

c. Hydrates.—Phosphoric acid forms three distinct hydrates; and in uniting with bases to form salts, it combines in different proportions, forming three classes of salts analogous to these hydrates.

aa. Protohydrate, metaphosphoric acid, glacial phosphoric acid, PO_5+HO; when it combines with bases, forms salts containing 1 equivalent of base, the latter replacing the 1 equivalent of water.

bb. Deutohydrate paraphosphoric acid, $PO_5 2HO$. Obtained by heating common phosphate of soda to redness, and decomposing the paraphosphate of soda then formed. This hydrate may be obtained by evaporating its aqueous solutions, in the form of imperfect crystals. It forms salts with 2 equivalents of base to 1 of acid, the base replacing the water.

cc. Terhydrate, common phosphoric acid, PO_5 $3HO$. Obtained by decomposing the common phosphate of soda; its solution evaporated, deposits crystals. When heated to redness, it is converted into the protohydrate; its salts contain 3 equivalents of base.

d. Phosphates of ammonia.

aa. Phosphate of ammonia, $2NH_3+3HO+PO_5$, or $2NH_4O+HO+PO_5$. A white, crystalline salt, formed direct, soluble in water, and decomposed by heat.

bb. Biphosphate of ammonia, $NH_3+3HO+PO_5$, or $NH_4O+2HO+PO_5$. Like the phosphate, formed direct, and decomposed by heat.

D. Phosphorus with chlorine.—Chlorides of phosphorus.

i. Protochloride, or sesquichloride, PCl_3. Formed when dry chlorine acts on excess of phosphorus; a transparent, colorless liquid; sp. gr. 1.45. Boils at 173°. It fumes in the air, has a suffocating odor, and is rapidly decomposed by water, hydrochloric and phosphorous acids being produced.

ii. Perchloride, PCl_5. Formed when chlorine in excess acts on phosphorus; the latter burns, and produces a white solid; volatile at 200°, fuming in the air, and decomposed by water, phosphoric and hydrochloric acids being formed. Boils at 299°. Both these chlorides absorb ammonia without being decomposed.

E. Phosphorus and iodine, bromine, $P\ I_3$ and $P\ I_5$, &c. They act strongly on each other, and form compounds analogous in composition to the chlorides.

F. Phosphorus and hydrogen.

i. Phosphuret of hydrogen, $P_2\ H$. A solid, yellow substance, insoluble in water, and burning at 320°, formed by throwing phosphuret of potassium into water. It becomes red when exposed to light.

ii. Liquid phosphuret of hydrogen, $P\ H_2$. Formed by exposing the phosphuretted hydrogen, evolved from phosphuret of calcium and water, to a freezing mixture; a colorless, transparent liquid, spontaneously inflammable in the highest degree; decomposed by light.

iii. Phosphuretted hydrogen gas, $P\ H_3$. Formed, when phosphorus is presented to nascent hydrogen, as when phosphorus is boiled in a solution of caustic potash; may be collected over water; a colorless, transparent gas, spontaneously inflammable in the air, in consequence of the presence of a small quantity of the vapor of $P\ H_2$; sp. gr. about 1.185. There is also a variety which does not inflame in the air, and that which is spontaneously inflammable soon loses that property when kept, particularly if exposed to light; inflammation is also prevented by the presence of minute quantities of certain vapors, as oil of turpentine, nitrous acid, &c.

G. Phosphorus and nitrogen, $P\ N_2$. Formed when the ammonia chloride of phosphorus is heated to redness out of contact of air; a white solid, decomposed if heated in the air, phosphoric acid being formed. It combines with water to form a hydrate, $P\ N_2 + 2\ H\ O$.

H. Phosphorus and sulphur.—Sulphuret of phosphorus,

P S. Formed direct by cautious fusion; a light yellow, semi-transparent liquid, not acted on by water.

5. **Selenium.** Symb. Se. Eq. 40. (39.57.) Discovered by Berzelius. Origin of name.

A. Exists combined with certain metals in sulphurets; very rare.

B. Obtained chiefly from a red deposit formed in the works at Gripsholm, where sulphuric acid is made from pyrites containing selenium.

C. Properties.—A dark reddish brown, brittle solid, having a semi-metallic lustre; sp. gr. 4.32; fusing at about 220°; boiling at 650°; its vapor has a peculiar odor, like decayed horseradish; transparent in thin layers; in many respects resembles sulphur. Solubility.

D. Selenium and oxygen.

i. Oxide of selenium, $Se\,O$. Formed during the imperfect combustion of selenium in air; a dark-colored solid.

ii. Selenious acid, $Se\,O_2$. Formed when selenium is burnt in oxygen; white, crystalline solid; deliquescent, soluble in water; combines with bases to form selenites.

iii. Selenic acid, $Se\,O_3$. Formed by the action of selenium in nitre, which produces seleniate of potash, and this, when decomposed, yields pure selenic acid. It has not been obtained anhydrous. The concentrated solution is colorless; sp. gr. 2.524; very sour; combines with bases, and forms seleniates. The acid is very deliquescent, and its solution is not altered by a heat of 530°.

E. Selenium and chlorine. — These are two chlorides, formed directly; $Se_2\,Cl$, a brown liquid heavier than water, and gradually decomposed by it; and $Se\,Cl_2$, a volatilizable, crystalline solid.

F. Selenium and iodine, &c. — Chloride and iodide of selenium may be formed direct.

G. Selenium and hydrogen. — Seleniuretted hydrogen. $Se\,H$. Obtained by the action of hydrochloric acid on seleniuret of potassium; a transparent, colorless gas; soluble in water; having a peculiar offensive odor, and highly poisonous.

H. Selenium with sulphur, and phosphorus.—These are solid compounds, which may be readily formed direct.

6. Boron. Symb. B. Eq. 11. (10.90.) Discovered simultaneously by Davy, Gay Lussac, and Thénard. Origin of name.

A. Exists native as boracic acid, pure in volcanic districts, and combined with soda and lime. Obtained from anhydrous boracic acid by ignition with potassium, which decomposes the acid, abstracting its oxygen.

B. Properties.—A dark, greenish-black solid, having neither taste nor smell; infusible; insoluble in water; sp. gr. 2·0; heated red hot, it burns in the air, and forms boracic acid.

C. Boron and oxygen.—Boracic acid, $B\ O_3$. The only known oxide, obtained from sassoline, the native boracic acid, by purification; or from the native borate of soda (Borax), by the action of sulphuric acid. Properties.—A white, crystalline solid; as deposited from its solution, it contains 3 equivalents of water, at a red heat this is expelled, and anhydrous acid remains; sp. gr. 1.8; slightly soluble in water; solution acid, reddens both litmus and turmeric, tinges flame green; combines with bases, and forms borates.

D. Boron and chlorine.—Chloride of boron, $B\ Cl_3$. Obtained by passing dry chlorine over heated boracic acid and charcoal; a transparent gas, decomposed by water, and having a sp. gr. of 3.942.

E. Boron and fluorine.—Fluoboric acid, $B\ Fl_3$. Formed by heating a mixture of dry boracic acid, fluoride of calcium, and sulphuric acid; a transparent, colorless gas; may be collected over mercury; instantly decomposed by water; fumes in the air, depositing boracic acid; highly pungent, irritating odor. Sp. gr. 2.37.

F. Boron and sulphur. A white solid, decomposed by water; may be formed direct. The other compounds of boron are unexamined, or of little interest.

7. SILICON. Symb. Si. Eq. 21. Discovered by Berzelius. Origin of name.

A. Exists native abundantly as an oxide, Silica.

B. Obtained by the action of potassium on silica, or on the silico-fluoride of potassium, and subsequent washing with water.

C. Properties.—A dark brown solid, infusible, having neither taste nor smell, burning in the air or oxygen, when strongly heated, and forming silica.

D. Silicon and oxygen.—Silex, silica, silicic acid, $Si\ O_3$.

i. Exists abundantly in nature, pure in rock crystal, quartz, flint, sand, tabasheer, &c.; and combined with lime, magnesia, alumina, &c., in most common stones.

ii. Obtained pure by decomposing an alkaline silicate by muriatic acid, washing and drying the precipitate.

iii. Properties.—Native silex, or that which has been ignited, is a white or colorless, tasteless, inodorous solid, insoluble in water, fusible at a white heat, and at lower temperatures with the alkalies and alkaline earths, &c., with which it forms salts, called silicates; sp. gr. 2.66. The precipitated silex is soluble in the strong mineral acids, in alkaline solutions, and to a small extent even in water. Used in making glass.

E. Silicon and chlorine.—Chloride of silicon, $Si\ Cl_3$. Formed direct, or by passing dry chlorine over an ignited mixture of silex and charcoal; a volatile, yellow liquid, instantly decomposed by water, gelatinous silica and hydrochloric acid being formed. Sp. gr. 1.52.

F. Silicon, iodine, and bromine; no iodide is known. The bromide is analogous to the chloride.

G. Silicon and fluorine.—Fluosilicic acid gas, $Si\ Fl_3$.

Formed by the action of hydrofluoric acid on silica, or by a mixture of fluoride of calcium, silica, and oil of vitriol; a transparent, colorless gas, which may be collected over mercury; sp. gr. 3.6. It fumes in the air, is strongly acid, and is instantly decomposed by water, hydrofluoric acid, gelatinous silica, and hydrofluosilicic acid being formed.

H. Silicon and sulphur, $Si\ S_3$. May be formed direct; a white solid, decomposed by continued ignition, or by the action of moisture.

8. Carbon. Symb. C. Eq. 6. Discoverer. Origin of name.

A. Exists native, pure, as the diamond, nearly pure in plumbago and some forms of stone coal; as carbonic acid, in the air, in limestone and various rocks; and with hydrogen, &c., in coal, and all forms of organic matter.

B. Obtained as charcoal by the imperfect combustion of wood and other forms of organic matter; pure, as lamp-black, from the imperfect combustion of highly carbonaceous substances.

C. Properties.—As diamond, very hard, brilliant, colorless, or nearly so, crystalline, transparent, solid, infusible, fixed, combustible at a white heat; as charcoal, black, softer, opaque, sometimes earthy-looking, sometimes brilliant metallic lustre, infusible, very unchangeable, combustible in the air at a red heat, or even below. Some forms of charcoal possess remarkable power of absorbing and condensing gases, in consequence of which they are used as purifiers and disinfectors.

D. Uses.—The diamond, as a gem; for cutting tools, as the glazier's diamond; as a cutting and grinding material; for pivot-rests in delicate machinery; as plumbago, to diminish friction; as charcoal, for fuel, as a deoxidizing agent, as a purifier, anti-putrescent, and filter for water, &c.

E. Compounds with oxygen.

i. Croconic acid, $C_5\ O_4 + H\ O$. Formed during the

preparations of potassium: by the action of heat on the salts of rhodizonic acid. A yellow crystalline solid; soluble in ether and alcohol, but slightly in water; sour; burning when heated.

ii. Carbonic oxide, $C\,O$. Obtained by heating chalk and zinc filings to redness, or by gently heating a mixture of ferrocyanide of potassium and sulphuric acid. Properties.—A gas, permanently elastic, transparent, colorless, tasteless, inodorous, poisonous; extinguishes flame, but burns itself, with a bright blue flame; may be collected over water; sp. gr. 0.9722.

iii. Rhodizonic acid, $C_7\,O_7 + 3\,H\,O$. Formed during the preparation of potassium by Brunner's method, and also by the action of carbonic oxide on heated potassium; obtained by decomposing the rhodizonate of potassium by sulphuric acid. A colorless, transparent, crystalline solid; combustible when heated; and combining with bases to form red crystalline salts. Soluble in water, alcohol, and ether.

iv. Oxalic acid, $C_2\,O_3 + H\,O$. Obtained most readily by acting on sugar by nitric acid and purifying by crystallization; exists native in rhubarb, sorrel, and many other plants; has not been obtained independent of water or a base. Properties.—A white, crystalline solid, subliming at about 300°, soluble in water, strongly acid; powerfully reddens vegetable blues; combines with bases to form oxalates; very poisonous.

a. Oxalic acid and ammonia.—Oxalate of ammonia, $C_2O_3 + NH_3 + 2\,HO$. A crystalline, white solid, readily formed direct, soluble in water, and useful as a test for the presence of lime in water, &c. There is also a binoxalate easily formed direct, $2\,C_2\,O_3 + N\,H_3 + 4\,H\,O$; and also a quadroxalate, $4\,C_2\,O_3 + NH_3 + 8\,HO$. When oxalate of ammonia is distilled, it loses the elements of 2 equivalents of water, and becomes oxamide, $C_2O_2H_2N$. There is also formed a portion of oxamic acid, $C_2H_4O_5N$. A white, granular solid; soluble in water; and combining with bases to form salts; its solution is decomposed by boiling.

v. Carbonic acid—fixed air, choke-damp—CO_2.

a. Exists in the atmosphere, formed by combustion, respiration, fermentation, and volcanic action; exists in the solid form in the native carbonates of lime, magnesia, iron, copper, &c.

b. Formed by burning any of the forms of carbon in the air or oxygen; by decomposing a carbonate by a more powerful acid.

c. Properties.—A transparent, colorless gas; sp. gr. 1.527; slightly soluble in water; a non-supporter of combustion or respiration; condensed to the liquid form by a pressure of about 40 atmospheres, and solidifiable by the cold produced during its own evaporation. Its acid powers are very feeble; it scarcely reddens vegetable blue tests. Combined with bases, it forms carbonates; and it is expelled with effervescence from these compounds by almost all other acids. At a high temperature, carbonic acid is decomposed by several of the metals, carbonic oxide being formed. Potassium and sodium decompose it entirely, carbon being separated. Carbonic acid is essential to the growth of plants. It is employed in the manufacture of various carbonates; and of effervescing mineral water.

d. Carbonic acid and ammonia.—Carbonates of ammonia.

aa. Carbonate of ammonia, $NH_3 + CO_2$. A white solid, formed direct, by mixing the dry gases together, and decomposed by water.

It is volatile at 140°, and smells strongly of ammonia.

bb. Sesquicarbonate of ammonia—salt of hartshorn—$2\,NH_3+3\,CO_2+2\,HO$. The common "carbonate of ammonia" of commerce, formed direct, or by heating sulphate or muriate of ammonia and chalk; a white, translucent salt, soluble in water, smelling strongly of ammonia, and gradually efflorescing in the air from the escape of ammonia; it has an alkaline reaction.

cc. Bicarbonate of ammonia, NH_3+2CO_2+2HO. Formed by the action of carbonic acid on the sesquicarbonate; a white, opaque solid, nearly inodorous and tasteless; soluble in water, and has a slightly alkaline reaction.

vi. Mellitic acids. $C_4\,O_3$, $H\,O$. Exists in the mineral *mellite* united with alumina. A colorless, crystalline solid; fusible; soluble in water and alcohol.

F. Carbon and chlorine.—Chlorides of carbon.

i. Dichloride of carbon.—Bicarburet of chlorine, $C_2\,Cl$. Formed by passing the vapor of the protochloride over red-hot porcelain, &c.; a transparent, colorless, crystalline solid, insoluble in water; combustible, burning with a red flame.

ii. Protochloride of carbon, $C\,Cl$. Formed by passing the vapor of the sesquichloride over ignited rock-crystal; a transparent, colorless liquid; sp. gr. 1.6; boils at about 248°; insoluble in water; will not burn alone, but does so in the flame of a lamp.

iii. Sesquichloride of carbon, $C\,Cl_3$. Formed by exposing a mixture of olefiant gas (carburetted hydrogen) and excess of chlorine to the influence of light; a transparent, colorless, crystalline solid, nearly tasteless, with an odor somewhat resembling that of camphor. It fuses at 320°, boils at 365°, and sublimes without change; sp. gr. about 2. It burns with difficulty, best in the flame of a lamp. It is insoluble in water.

iv. Bichloride of carbon, $C\,Cl_2$. Formed by passing chlorine and vapor of sulphuret of carbon through a red-hot porcelain tube into a cooled condenser, and subsequently agitating the liquid obtained with a solution of caustic potash, which removes the chloride of sulphur thus formed, and leaves the chloride of carbon; a colorless liquid, having a suffocating odor; sp. gr. 1.59; boils at 172°; insoluble in water.

G. Carbon and iodine.—Iodides of carbon.

i. Protiodide of carbon, $C\,I$. Formed by the mutual action of dry sesquiodide of carbon and perchloride of phosphorus; a heavy yellow, oily liquid, having a sweet taste and ethereal odor.

ii. Sesquiodide of carbon, $C_2\,I_3$. Formed when strong alcoholic solutions of iodine and potassa are mixed; a lemon yellow crystalline solid; sp. gr. nearly 2.0; insoluble in water; fuses at 250°; volatile at lower temperatures.

H. Bromides of carbon. A protobromide, $C\,Br$, is formed by acting on alcoholic solution of potash by bromine; a white volatile solid, insoluble in water, and having a fragrant odor. A sesquibromide, $C_2\,Br_3$, is obtained by acting on the iodide of carbon by bromine. It is a colorless liquid, volatile, and insoluble in water.

I. Carbon and hydrogen.—Hydrocarbons.

i. Bihydroguret of carbon, light carburetted hydrogen, $C\,H_2$. Exists native, as fire-damp in coal-mines, and the inflammable air of marshes; artificially obtained by heating a mixture of acetate of soda with caustic lime and potassa; a colorless transparent gas, very slightly soluble in water; sp. gr. 0.555; hence 100 cubic inches weigh 17.4 grains; extinguishes flame, but burns itself with a pale yellow flame; forms an explosive mixture with common air; 8 volumes of air to 1 of gas most explosive.

ii. Methyle, $C_2\,H_3$. A hypothetical compound. (See p. 61.)

iii. Ethyle, $C_4 H_5$. A hypothetical compound. (See p. 54.)

iv. Amyle, $C_{12} H_{11}$. A hypothetical hydrocarbon. (See p. 64.)

v. Carburet of hydrogen, olefiant gas, elayle, $C_2 H_2$. An organic radical. Obtained by heating a mixture of 1 volume of alcohol and 2 volumes of oil of vitriol; a colorless gas, which by cold and pressure may be liquefied; it may be collected over water; its sp. gr. is 0.98; 100 cubic inches weigh 31.7 grains; it extinguishes flame, but burns with a bright and smoky flame; when pure it has neither taste nor smell, and is without action on test-papers.

a. Olefiant gas and chlorine.—Chloride of elayle, $C_2 H_2 Cl$. Formed when olefiant gas is mixed with twice its volume of chlorine, condensation takes place; a transparent colorless liquid; sp. gr. 1.2; it boils at 152°, and burns with a green and smoky flame; exposed to the action of excess of chlorine and light, chloride of carbon is formed.

b. Olefiant gas and iodine.—Iodide of elayle, $C_2 H_2 I$. Formed like the chloride; a white or yellowish crystalline solid; fusible at 174°; insoluble in water; sp. gr. 1.84.

c. Olefiant gas and bromine.—Bromide of elayle, $C_2 H_2 Br$. Formed direct, or by the action of bromine on the iodide; a colorless liquid; heavier than water, and insoluble in it. Boils at 265°.

vi. Quadrihydrocarburet, etherine, $C_4 H_4$. Formed during the destructive distillation of fat oils; a transparent colorless gas; sp. gr. 1.963; condensed at 0° to a liquid; very combustible; burning with a bright and smoky flame.

vii. Bicarburet of hydrogen, $C_6 H_3$. Formed during the decomposition of fat oils by heat; a transparent colorless liquid; sp. gr. 0.85; boils at 186°;

very combustible, and burning with a bright and smoky flame.

viii. Cetene, $C_{32} H_{32}$. Obtained by distilling ethal with pure phosphoric acid; a volatile oily liquid, which boils at 528°, and burns with a pure bright flame; its hydrate, ethal, $C_{32} H_{32} + 2 H O$, is obtained by the saponification of spermaceti, which converts cetine into ethal. Ethal is a white solid fat; fusible at 118°, and soluble at a higher temperature. Cetine, the purified fat of spermaceti, consists of $C_{208} H_{208} O_{16}$, and fuses at 120°.

ix. Amilen, $C_{10} H_{10}$. Obtained by distilling potato oil (the hydrated oxide of amyle, see p. 64) with phosphoric acid; it is a transparent, colorless liquid, which boils at 320°. Besides these, several other similar hydrocarbons, likewise isomeric with olefiant gas and etherine, have been described.

x. Caoutchouc, India rubber, gutta percha, $C_8 H_7$.

a. Caoutchouc exists naturally in the juices of many plants, as *Ficus*, *Siphonia*, *Urceola*, &c. Obtained by the inspissation of the exuded juice; a dark-grayish brown, translucent solid; very elastic; insoluble in water; very combustible, burning with a smoky flame; soluble in naphtha, caoutchoucine, ether, camphene, &c. When distilled, caoutchouc yields a mixture of several hydrocarbons, termed caoutchoucine; a volatile, transparent, colorless liquid, which contains etherine, hevæne, &c. Caoutchouc is acted on in a remarkable manner by sulphur, absorbing a small portion of that substance when it is said to be vulcanized; it is then tougher, far more elastic, less affected by heat, and less easily dissolved than ordinary caoutchouc.

b. Gutta percha. Obtained, like caoutchouc,

from the *Isonandra Gutta;* a brown, somewhat elastic, very tough solid; sp. gr. about 0.979; becomes soft, plastic, and very adhesive at 150°, hardening again as it cools; melts at 250°, being partially decomposed, very inflammable; soluble in ether, oil of turpentine, chloroform; may be perfectly blanched.

xi. Naphtha, $C_6 H_5$. Exists native in Italy, Persia, India, &c.; obtained artificially during the destructive distillation of coal. Properties.—A colorless limpid liquid; sp. gr. varies from 0.750 to 0.800; boils at from 320° to 360°; very inflammable, and burns with a bright smoky flame; dissolves most resins and oils, &c.; used as a solvent for caoutchouc, and to preserve potassium and similar very oxidizable substances.

xii. Acetyle, $C_4 H_3$. A hypothetical hydrocarbon. (See p. 59.)

xiii. Camphene, oil of turpentine, $C_{10}H_8$. Exists naturally in the resinous exudation of coniferous trees; obtained by distillation and rectification; a limpid colorless liquid; sp. gr. 0.87; boils at 314°; insoluble in water; very inflammable; burns with a very bright smoky flame. The oil of lemons (citrene), of orange peel, of bergamot, of orange flowers, of sabine, of parsley, of pepper, of cubebs, &c., are isomeric in composition with camphene, and though differing from it in odor, resemble it in most other characters.

a. Camphene and oxygen.—Camphor, $C_{10} H_8 O$. There is obtained from the camphor tree a volatile oil, isomeric with camphene ($C_{10}H_8$); and this by oxidation forms camphor and a hydrate of camphene. Camphor is a white, semi-transparent, volatile, crystalline, tough yet brittle solid, having an aromatic odor and bitter taste; sp. gr. 0.987; it fuses at 347°; boils at 400°; and is easily combustible. Camphor absorbs muriatic acid to form a fluid hydrochlorate; by the action of hot nitric acid it is converted into camphoric acid, $C_{10} H_7 O_3$ which forms salts, and crystallizes with H O. When camphor is distilled with anhydrous phosphoric acid it is converted into camphone, $C_{10} H_7$ (?).

b. Camphene and water. — Hydrate of camphene, $C_{10} H_8 + 2 HO$, or with 3 H O. Formed when turpentine and water are long kept in contact; white crystalline solids, resembling camphor. A hydrate of camphene, $C_{10} H_8 + H O$, is also produced by certain trees, constituting Borneo camphor. It may be formed artificially from the volatile oil of camphor by boiling it with potassa.

xiv. Colophene, $C_{20} H_{16}$. Formed during the distillation of oil of turpentine with sulphuric acid; a colorless liquid seen by transmitted light, but bluish by partly reflected light; sp. gr. 0.940; boiling point about 600°.

xv. Sesquihydrocarburet, $C_3 H_2$. Formed with the bicarburet in the destructive distillation of oil; a transparent liquid.

xvi. Formyle, $C_2 H$. A hypothecal hydrocarbon. (See p. 63.)

xvii. Scheererite, koenleinite, $C_{32} H_{16}$. A soft, light, greasy-looking, combustible solid; found native in brown coal.

xviii. Naphthaline, $C_{10} H_4$. Formed during the destructive distillation of coal, oil, &c.; a white crystalline solid, having a peculiar aromatic odor; sp. gr. 1.05; insoluble in water; slowly volatile at common temperatures; fuses at 176°; boils at about 420°; very inflammable; burns with a very smoky flame. Associated with naphthaline, and, like it, formed during the destructive distillation of coal, oil, &c., is an isomeric hydrocarbon called paranaphthaline, $C_{15} H_6$. It fuses at 356° ; boils at about 570° ; and may be obtained crystallized by sublimation.

a. Naphthaline and chlorine. By the action of chlorine two compounds are formed; a protochloride, $C_{10} H_4 Cl$, a yellow oily liquid; and a bichloride, $C_{10} H_4 Cl_2$, a white crystalline solid, fusing at 320°.

b. Naphthaline and sulphuric acid.—Sulphonaphthalic acid, $2C_{10} H_4 + 2 S O_3$. Formed direct when the two are heated together; a white crystalline solid; fusible at 212°; soluble in water; acid; and combining with bases to form sulphonaphthalates.

xix. Benzoyle, $C_{14} H_5$. A hypothetical hydrocarbon. (See p. 66.)

xx. Tar series. These are natural or artificial mixtures of different hydrocarbons and resinous substances; their exact composition is not well made out. Wood tar has been found to contain, or yield when carefully distilled, Creosote, or kreosote, a compound of C H and O, the exact proportions being unknown; it is a colorless oily liquid; sp. gr. 1.037; boils at 397° ; burns with difficulty, and is remarkable for its antiseptic powers. Paraffine, a colorless crystalline solid, said to consist of $C_{24} H_{25}$; it fuses at 110°, volatilizes, and burns with a bright though smoky flame. Eupione, a colorless liquid; sp. gr. 0.655; said to contain $C_5 H_6$; it boils at 116°, and readily burns. Kapnomor, a colorless oily liquid; sp. gr. 0.977; boils at 365°, and does not congeal at —6°. Picamar, a colorless, greasy, intensely bitter liquid; sp. gr. 1.10; boils at 545°.

xxi. Coal gas, oil gas, resin gas. These gases, artificially obtained by the destructive distillation of coal, oil, and resin, consist of mixtures of some of the preceding hydrocarbons.

a. Manufacture of coal gas.—Destructive distillation of bituminous coal in ignited iron cylinders; separation of tar and water containing salts of ammonia in the "hydraulic main," passage of gas through condenser or refrigerator, for further condensation of water and other vapors; purification in the lime purifiers, either dry or wet lime; difficulty of removing sulphur from gas; sp. gr. of coal gas, 0.450 to 0.650.

b. Combustion of gas.—Nature of flame; contact of the combustible and the supporters of combustion, at a certain temperature required; heat of flame depends on the perfect chemical action of the two; its light depends on the ignition of solid matter; flame hollow; effect of cooling flame; safety lamp; temperature necessary for flame; perfect combustion; the Argand lamp; imperfect combustion; want of air or heat; formation of smoke; modes of increasing heat; the blowpipe; bellows, and blowing machines; oxidators; use of oxygen; modes of increasing light; the Drummond light.

J. Carbon and nitrogen.—Cyanogen, $N C_2$.

i. Exists in combination in certain vegetable products; obtained by heating the cyanide of mercury. Properties.—A transparent colorless gas; absorbed by water, but may be collected over mercury; condensed to the liquid state by a pressure of four atmospheres, and solidifiable at —30° to a transparent crystalline solid. The gas has a penetrating and peculiar odor, like that of bitter almonds; is highly poisonous; extinguishes flame, but itself burns with a beautiful purple-red flame; sp. gr. 1.82. Cyanogen, though a compound, has a remarkable analogy to chlorine, and some other elements.

ii. Paracyanogen. The residue of the preparation of cyanogen from the cyanide of mercury is a brown solid, neither fusible nor volatile; this appears to consist of C. N.

iii. Mellone, $C_6 N_4$. Formed during the destructive distillation of sulphocyanogen; a yellow solid; insoluble in water, fixed at a low red heat, which combines with the metals to form mellonides, and with hydrogen to form a hydromellonic acid, $C_6 N_4 H$.

iv. Cyanogen and oxygen.

a. Cyanic acid, $C_2 N + O$. Obtained by the distillation of cyanuric acid, when hydrated cyanic acid is formed. This is a strong acid liquid; very prone to change; the acid has not been obtained, hitherto, independent of water or some base. It unites to bases to form cyanates.

aa. Cyanic acid and ammonia.—Cyanate of ammonia. When ammonia and hydrated cyanic acid are brought together, they combine and form a white solid salt, $C_2 N O + 2 N H_3 + H O$; this, when gently heated, becomes cyanate of ammonia, $C_2 N O + 2 N H_3 + H O$, or Urea. This substance, which always exists in urine, being formed in the animal system, may be regarded either as a cyanate of ammonia, as an independent compound represented by $C_2 H_4 O_2 N_2$, or as an amide of carbonic oxide, $2 N H_2 + C_2 O_2$. It is obtained from urine by cautious evaporation, and the addition of nitric acid, which forms a crystalline nitrate of urea.

b. Paracyanic acid, fulminic acid, $2 C_2 N + O_2$. This acid has only been obtained in combination with bases; when set free from its compounds it always suffers decomposition; its salts, called fulminates, are easily decomposed, and remarkable for the violence with which most of them explode when suddenly decomposed by heat, friction, &c.

c. Cyanuric acid, $3\ C_2N + O_3$. Formed when urea is decomposed by heat; a white solid; it can only be obtained in combination or hydrated with 3 or 4 equivalents of water; it combines to form cyanurates.

v. Cyanogen and chlorine.—Chlorides of cyanogen.

a. Protochloride of cyanogen, $C_2N\,Cl$. Formed when hydrocyanic acid is acted on by chlorine, or by exposing cyanide of mercury to the action of moist chlorine and sunshine; a gas easily condensible by cold, into a liquid or solid; it has a very powerful pungent odor, and is highly poisonous.

b. Solid chloride of cyanogen, $2\ C_2N + 2\ Cl$. Formed when dry chlorine and hydrocyanic acid are exposed to sunshine; a white crystalline solid; sp. gr. 1.32; fuses at 284°; sublimes at 374°; decomposed by water; very poisonous.

vi. Cyanogen and iodine.—Iodide of cyanogen, $C_2N + I$. Formed by heating a mixture of cyanide of mercury and iodine; a white crystalline solid; soluble in water; having a very pungent odor; highly poisonous.

vii. Cyanogen and bromine, $C_2N + Br$. Analogous to the iodide in most points.

viii. Cyanogen and hydrogen.—Hydrocyanic or prussic acid, $C_2N + H$. Exists naturally in some plants; obtained artificially by distilling cyanide of mercury with hydrochloric acid, or by decomposing it by sulphuretted hydrogen, and condensing the vapor by a cold of 0°, &c.; a transparent, colorless liquid, very volatile, and easily solidifying by the cold of its own evaporation; having a strong and penetrating odor like crushed laurel leaves; highly poisonous; sp. gr. 0.700; it boils at 80° ; combustible, burning with a bright flame ; very prone to decompose, which is prevented by the presence of a trace of other acids. Its acid powers are very feeble ; it does not expel carbonic acid from the carbonates. It readily dissolves in water ; a dilute solution is employed in medicine.

ix. Cyanogen and sulphur.—Sulphocyanogen (bisulphuret of cyanogen), $C_2 N S_2$. Formed when sulphur and cyanide of mercury are heated together; a yellow solid, insoluble in water; combines with the metals to form sulphocyanides.

a. Sulphocyanogen and hydrogen.—Hydrosulphocyanic acid, $C_2 N + S_2 + H$. Occurs in the seeds and blossoms of the *Cruciferæ* and in the saliva of man and sheep. Obtained by decomposing sulphocyanide of lead by sulphuric acid. It has not been obtained anhydrous. A colorless fluid. It unites to bases, and forms hydrosulphocyanates or sulphocyanides.

x. Cyanogen and selenium.—Seleniocyanogen, $N C_2 Se_2$. Exists in combination.

K. Carbon and sulphur. — Bisulphuret of carbon, $C S_2$. Obtained by passing the vapor of sulphur over ignited charcoal, and condensing the result by cold ; a transparent, colorless liquid, insoluble in water, possessing a strong and disagreeable fetid odor; sp. gr. 1.272 ; boils at 118° ; very volatile ; dissolves sulphur, phosphorus, most oils, resins, &c. ; exists often in coal gas, and cannot be separated by lime like the other sulphur compounds in coal gas.

L. Carbon and phosphorus. — Phosphuret of carbon. Formed by acting on phosphuret of calcium by water, and washing the residue with hydrochloric acid; a yellow, tasteless, odorless solid, burning at a red heat.

Section IV.—Compounds of Carbon, Oxygen, Hydrogen, and Nitrogen, etc. Organic Matter.

1. Starch.

A. Common starch, or fecula, $C_{12} H_{10} O_{10}$. An important element of animal food. Exists in nearly all plants; separated from the other substances with which it is associated by mechanical means. General properties.—A white, granular, pulverulent solid, insoluble in cold water; tasteless, odorless; soluble in water at 158°; sp. gr. 1.50; acquires a deep blue by the action of iodine. Nitric acid converts it into a highly combustible substance, xyloidine, consisting of $C_{15} H_{12} O_{16} N$. The chief sources of starch are wheat, potatoes, rice, arrowroot, sago palm, &c. The different kinds of starch vary slightly in chemical composition, the size of the grains, &c. Action of Diastase.

B. Inulin, $C_{24} H_{21} O_{21}$. A variety of starch which is found in the dahlia, and many other plants; sp. gr. 1.356; fuses at 212°; rendered yellow by iodine. Its aqueous solution does not gelatinize on cooling.

C. Lichenin, $C_{24} H_{20} O_{20}$. The starch of lichens: iodine turns it green. There are also a number of other varieties of starch to which different names are given in relation to the plants from which they are obtained.

2. Gum.

A. Arabine, gum Arabic, $C_{12} H_{11} O_{11}$. Exists in many plants, and is obtained in the form of a natural exudation; a brittle, translucent, white or pale yellowish solid; odorless; nearly tasteless; soluble in water, with which it forms mucilage; insoluble in alcohol; precipitated white by subacetate of lead. By the action of nitric acid it is converted into mucic acid.

B. Gum tragacanth. Exists in many plants; differs from arabine in not being soluble in cold water.

C. Dextrine, or British gum, $C_{24} H_{20} O_{20}$. Formed when starch is roasted or exposed to a temperature of nearly 400°, or when it is boiled with dilute sulphuric acid, or by the action of diastase, a substance found in germinating seeds, and which converts the starch in them into dextrine; a white or pale buff solid, soluble in cold water, and closely analogous to gum.

D. Pectine, or pectic acid, $C_{12} H_8 O_{10}$. A substance analogous to gum, which exists in many plants, especially in the fruits. It is a tasteless solid, which swells up and gelatinizes with water. It possesses feeble acid powers, combining with bases to form pectates.

E. Mucic acid, saclactic acid, $C_{12} H_8 O_{14} + 2 HO$. Obtained as a hydrate by the action of nitric acid on gum; a white granular solid, slightly soluble in water; feebly acid; unites to bases to form mucates or saclactates.

i. Mucic acid and ammonia.—Mucate of ammonia, $C_{12} H_8 O_{14} + 2 N H_3 + H O$. A white crystalline salt, formed direct, and soluble in water. Is permanent at 212°.

F. Pyromucic acid, $C_{10} H_3 O_5 + HO$. Formed by the destructive distillation of mucic acid; a white crystalline solid; very sour; soluble in water; fuses at 270°, and combines with bases to form pyromucates.

3. Lignin. $C_{36} H_{24} O_{24}$.

Exists in most plants, constituting their fibre or woody matter; nearly pure in flax and well-dressed hemp; white fibrous solid; sp. gr. about 1.5; tasteless, inodorous; insoluble in water; readily combustible at a dull red heat. Digested in strong nitric acid, it is converted into xylodine, of which "gun cotton" is one of the most perfect forms. Boiled with nitric acid, it forms oxalic acid; boiled with dilute sulphuric acid, it passes into gum. Pure lignin is not prone to decay, but, associated with other organic substances, it frequently decomposes rapidly. Various means are adopted to prevent this: corrosive sublimate and various other metallic salts are used for this purpose. Lignin has a strong affinity for several earthy and metallic compounds, and also for various forms of coloring matter: upon this depend the arts of dyeing and calico printing. Submitted to destructive distillation, lignin yields tar, water, wood spirit, acetic acid, &c. During the mouldering of lignin, it evolves water and carbonic acid, becoming brown, and passing into humus; when this change is complete it constitutes mould. These substances exist abundantly in peat, brown coal, &c. Most fertile soils contain more or less humus. Humus or humic acid, $C_{40} H_{12} O_{12}$, possesses feeble acid powers, and combines with bases to form salts.

4. SUGAR.

A. Cane or crystallizable sugar, $C_{12} H_{11} O_{11}$. Exists in many plants, and is obtained from their juices by evaporation; white crystalline solid; inodorous, having a sweet taste; soluble in water; sp. gr. about 1.55; fusible at 300°, when it becomes caramel, and is slightly modified, having lost 2 equivalents of water, and become $C_{12} H_9 O_9$. Boiled with dilute sulphuric or hydrochloric acid, cane sugar is converted into glucose. Nitric acid converts it into oxalic acid, with the evolution of copious nitrous fumes. Sugar combines with bases to form definite compounds, termed saccharides.

B. Glucose, granular, or grape sugar, $C_{12} H_{14} O_{14}$. Exists native in various plants, and readily formed from cane sugar, or from gum and starch, by the action of dilute acids, or diastase; or from lignin by the action of oil of vitriol; an imperfectly crystalline granular solid; inodorous; less soluble in water than cane sugar, but rather sweeter. Like cane sugar, it combines with bases and forms glucosides.

Glucic acid. Saccharic acid.

C. Mannite, $C_6 H_7 O_6$. A variety of sugar which exists in many plants, as the ash, larch, beet, &c.

D. Honey. Collected by bees from various flowers; consists of glucose and an uncrystallizable sugar. Besides these there are several other varieties of sugar, as that of liquorice, of fungi, of the oak, &c.

E. Lactine, sugar of milk, $C_{12} H_{10} O_{10} + 2 H O$. Exists in the whey of milk, and obtained from it by evaporation; a white crystalline solid; sp. gr. 1.543; fusible at about 300°, when it becomes anhydrous. It is less soluble in water than cane sugar, and much less sweet to the taste. Boiled with dilute acids it is converted into glucose.

i. Lactic acid, $C_6 H_5 O_5$. Formed by the fermentation of lactine; exists in sour milk, and in the flesh of animals and in the fermented juice of certain vegetables. A colorless syrupy liquid, having a strong affinity for moisture and an acid taste. Sp. gr. 1.215 at 68°. It unites to bases to form lactates, and, when distilled, yields, amongst other products, pyrolactic acid, $C_6 H_4 O_4$, a white crystalline solid, which fuses at 225°, boils at 480°, and combines with bases.

5. Ether Series.

A. Alcoholic fermentation. When grape sugar or glucose ($C_{12} H_{12} O_{12} + 2 H O$) is kept in contact with decomposing azotized matter, it ferments, or undergoes decomposition; by example 4 equivalents of carbonic acid and 2 of water are evolved, and 2 equivalents of alcohol ($C_4 H_6 O_2$) are left. Under the same circumstances cane sugar also ferments, passing first, however, into the intermediate state of glucose. Upon this change depends the formation of spirit in wine, brandy, beer, and all fermented liquors. An important series of compounds exist, formed by the decomposition and transformations of alcohol. These are all traced to a common radical, called ethyle, of which ether is considered the oxide.

B. Ethyle, $C_4 H_5$. Not obtained separate; a hypothetical radical.

C. Oxide of ethyle, ether, sulphuric ether, $C_4 H_5 O$. Formed by distilling a mixture of alcohol and sulphuric acid; a highly volatile, transparent, colorless liquid; sp. gr. 0.715; having a peculiar, fragrant odor, and a pungent, sweetish taste. It does not congeal at —166°. Boils at 96°. Its vapor is very inflammable, burning with a bright flame. It is somewhat soluble in water, which dissolves about one-ninth of it. It mixes in all proportions with alcohol, and dissolves sulphur, phosphorus, iodine, caoutchouc, and nearly all resinous and oily substances; causes insensibility when its vapor is inspired. Letheon.

i. Oxide of ethyle and water.—Hydrated oxide of ethyle, alcohol, $C_4 H_5 O + H O$. Well known under the form of "ardent spirits." Formed by the fermentation of glucose, or of any substance capable of being converted into glucose, as starch, &c.; separated and obtained pure by distillation and rectification; a transparent, colorless liquid, having a weak, but agreeable odor, and a strong, pungent taste; sp. gr. 0.794, boiling at 173°. It has a strong affinity for water, mixing with, and dissolving in it, in all proportions. Alcohol dissolves most resins, and is in consequence much used in the manufacture of resinous varnishes. Alcohol may be regarded as hydrated oxide of ethyle, $C_4 H_5 O + H O$, or as a binhydrate of etherine, $C_4 H_4 + {}_2H O$. The oxide of ethyle, or ether, combines with most acids to form compounds analogous to ordinary salts: these are termed ethers, and are of course very numerous.

ii. Nitrous ether, nitrite of oxide of ethyle, $C_4 H_5 O + N O_3$. Formed by the action of nitric acid on alcohol; a yellowish liquid, having an agreeable aromatic odor, sp. gr. 0.947; boiling at 70°; very inflammable.

iii. Nitric ether, nitrate of oxide of ethyle, $C_4 H_5 O +$

NO_5. Obtained by distilling a mixture of nitric acid and alcohol with nitrate of urea; a transparent, colorless liquid, having an aromatic odor, a sweet and slightly bitter taste; boils at 185°; sp. gr. 1.112; insoluble in water, but readidy soluble in alcohol.

iv. Sulphurous ether, sulphite of oxide of ethyle, $C_4H_5O + SO_2$. Obtained by distilling a mixture of alcohol and chloride of sulphur; a colorless liquid; sp. gr. 1.085; boiling at 322°; it has a mint-like odor, a hot sulphureous taste, and dissolves in ether and alcohol.

v. Sulphatic ether, oil of wine, sulphate of oxide of ethyle, $C_4H_5O + SO_3$. Formed during the distillation of alcohol and sulphuric acid, and comes over towards the end of the process; a yellow, aromatic, oily fluid; soluble in alcohol and ether, insoluble in water; sp. gr. 1.133.

vi. Sulphovinic acid, hydrated sulphate of oxide of ethyle, $C_4H_5O + 2SO_3 + HO$. Formed when sulphuric acid and alcohol are mixed, and immediately decomposed when the mixture is heated, as in the process for ether; by forming a sulphovinate of lead, and subsequently decomposing it by sulphuretted hydrogen, a solution of sulphovinic acid may be obtained; it exists as a sour liquid; sp. gr. 1.319; soluble in water, decomposed by heat, and combining with bases to form salts.

a. Sulphovinic acid and ammonia.—Sulphovinate of ammonia, $NH_3 + HO + C_4H_5O + 2SO_3$. Formed direct; a white crystalline solid; soluble in water, alcohol, and ether, and fuses at 144°.

vii. Ethionic acid, bisulphate of oxide of ethyle, $C_4H_5O + 2SO_3$. Formed when the vapor of anhydrous sulphuric acid is passed into absolute alcohol; combines with bases to form salts.

viii. Etherophosphorous acid, hydrated phosphite of

oxide of ethyle, $C_4H_5O + 2HO + PO_3$. Formed by the action of chloride of phosphorus on alcohol.

ix. Etherophosphoric acid, phosphovinic acid, hydrated phosphate of oxide of ethyle, $C_4H_5O + 2HO + PO_5$. Formed during the action of phosphoric acid on alcohol; it combines with bases and forms etherophosphates. Syrupy and colorless.

x. Boric ether, borate of oxide of ethyle, $C_4H_5O + 2BO_3$. Formed by the action of dry boracic acid on alcohol; a transparent, viscid, glassy solid.

xi. Silicic ether, silicate of oxide of ethyle, $C_4H_5O + SiO_3$. Formed during the action of chloride of silicon and alcohol; a colorless transparent liquid; sp. gr. 1.035; boiling at 572°; soluble in ether and alcohol; insoluble in water; combustible; burns with a white flame and leaves silica; there is also a basic silicate, $2C_4H_5O + SiO_3$.

xii. Oxalic ether, oxalate of oxide of ethyle, $C_4H_5O + C_2O_3$. Formed when alcohol, oxalic acid, and sulphuric acid are distilled together; an oily liquid; sp. gr. 1.0929; boils at 360°; it combines with ammonia to form a white, fusible, solid substance.

a. Oxalovinic acid, hydrated oxalate of oxide of ethyle, $C_4H_5O + HO + 2C_2O_3$. Formed by the action of bases on oxalic ether; combines with bases.

xiii. Carbonic ether, carbonate of oxide of ethyle, $C_4H_5O + CO_2$. Formed by the action of potassium on oxalic ether; a limpid, colorless liquid; there is also a hydrated carbonate, $C_4H_5O + HO + 2CO_2$

D. Chloride of ethyle, muriatic ether, C_4H_5Cl. Obtained by distilling a solution of muriatic acid in alcohol; a limpid, colorless liquid; sp. gr. 0.874; boils at 60°; solidifies at —10°; soluble in 50 parts of water, and in all proportions in ether and alcohol.

E. Iodide of ethyle, hydriodic ether, $C_4 H_5 I$. **Formed** by distilling a mixture of alcohol, iodine, and phosphorus; a transparent, colorless liquid; sp. gr. 1.92; boils at 148°; is not inflammable.

F. Bromide of ethyle, hydrobromic ether, $C_4 H_5 Br$. Formed like the iodide; a volatile liquid; sp. gr. 1.45.

G. Fluoride of ethyle, hydrofluoric ether, $C_4 H_5 F$. Formed like the chloride; a colorless liquid, very volatile, and combustible. Has the odor of horseradish.

H. Sulphuret of ethyle, $C_4 H_5 S$. Obtained by the action of chloride of ethyle on alcohol and protosulphuret of potassium; a transparent, colorless liquid; sp. gr. 0.825; boils at 153°; insoluble in water.

i. Hydrosulphuret of sulphuret of ethyle, hydrosulphuric ether, mercaptan, $C_4 H_5 S + H S$. Obtained by distilling a mixture of hydrosulphuretted sulphuret of potassium with sulphovinate of potassa; a colorless liquid, having a taste and odor resembling garlic; sp. gr. 0.842; boils at 97°; sparingly soluble in alcohol and water; combines direct with the metals to form compounds termed mercaptides; mercaptan may be represented as alcohol, the oxygen of which is replaced by sulphur.

I. Cyanide of ethyle, hydrocyanic ether, $C_4 H_5 Cy$. Obtained by distilling cyanide of potassium with sulphovinate of baryta; a colorless, poisonous liquid; sp. gr. .700; boils at 180°; slightly soluble in water; easily in ether and alcohol.

K. Sulphocyanide of ethyle, sulphocyanic ether, $C_4 H_5 + Cy S_2$. Formed when sulphocyanide of potassium, alcohol, and sulphuric acid are distilled together; a colorless liquid; sp. gr. 1.02; boils at 295°.

6. ACETYLE SERIES.

A. Acetic fermentation. When liquids, in which the alcoholic fermentation has taken place, are kept exposed to air, under conditions favorable to fermentation, a new kind of fermentation commences, under the influence of which alcohol passes into acetic acid. Alcohol, $C_4H_6O_2$, by the action of oxygen, loses 3 equivalents of hydrogen, which form water, and acquires an additional equivalent of oxygen, becoming acetic acid, $C_4H_3O_3$. This acid, and the various substances connected with it, are supposed to be derived from acetyle.

B. Acetyle, C_4H_3. A hypothetical base, corresponding to the ethyle of the ether series.

C. Oxide of acetyle, C_4H_3O. A hypothetical oxide, corresponding to ether.

i. Hydrated oxide of acetyle, aldehyde, $C_4H_3O + HO$. Obtained by distilling a mixture of alcohol, sulphuric acid, and black oxide of manganese; a colorless volatile liquid; sp. gr. 0.79; boils at 72°; has an ethereal, yet suffocating odor; readily soluble in ether, alcohol, and water; combines with ammonia to form a colorless, crystalline solid; fusible at 165° and volatile at 212° ($C_4H_3O + HO + NH_3 + HO$). Thialdine.

ii. Deutoxide of acetyle, aldehydic acid, $C_4H_3O_2$. This acid has not been satisfactorily examined; it appears to be formed during the imperfect combustion of ether.

iii. Peroxide of acetyle, acetic acid, pyroligneous acid, pure vinegar, $C_4H_3O_3 + HO$. The result of the acetous fermentation; obtained by the fermentation of vinous or spiritous liquors, and by the destructive distillation of wood; the pure acid is prepared from good vinegar by combining it with a base, and decomposing the purified salt thus formed by sulphuric acid. It has not been obtained separate in the anhydrous state. The hydrate is a transparent, colorless liquid, which freezes at about 45°, boils at 248°; the sp. gr. is 1.063; its odor is sour, and highly pungent; it readily combines with bases to form acetates; its vapor is combustible, burning with a pale bluish flame. Under the influence of light, chlorine forms a chloracetic acid, a colorless and very deliquescent solid, which fuses at 115°, boils at 390°, has a sp. gr. of 1.617, combines with bases, and consists of $C_4\ Cl_3\ O_3 + H\ O$. It is therefore acetic acid in which the hydrogen is replaced by chlorine; and with anhydrous sulphuric acid it forms a sulphacetic acid. In the manufacture of vinegar from wood, the crude pyroligneous acid obtained is much contaminated with tar, &c.; it is therefore neutralized with lime, the acetate of lime thus formed decomposed by sulphate of soda, and the resulting acetate of soda distilled with sulphuric acid.

a. Acetic acid and ammonia.—Acetate of ammonia, $C_4\ H_3\ O_3 + H\ O + N\ H_3$. Formed direct; a white, crystalline, and very deliquescent salt. There is also a binacetate.

iv. Acetone, pyroacetic spirit, $C_3\ H_3\ O$. Formed when dry acetates are decomposed by heat; a colorless limpid liquid; sp. gr. 0.792; boils at 132°; very inflammable, burning with a bright flame; readily miscible with water, ether, and alcohol. Chlorine expels 1 equivalent of hydrogen, and forms an oily fluid, consisting of $C_3\ H_2\ Cl\ O$.

v. Metacetone, $C_6\ H_5\ O$. Formed in certain cases along with acetone, and regarded as the oxide of a hypothetical base named mesityle, or metacetyle; a colorless liquid; insoluble in water; soluble in ether and alcohol; boils at 183°. When distilled with sulphuric acid and bichromate of potash it forms metacetonic acid, $C_6\ H_5\ O_3 + H\ O$; a volatile acid, somewhat resembling the acetic.

D. Acetyle and chlorine.—Terchloride of acetyle, $C_4\ H_3\ Cl_3$. Formed when chloride of ethyle and chlorine are exposed to the influence of sunlight. It may be regarded as acetic acid, in which the oxygen is replaced by chlorine, just as in chloracetic acid the hydrogen is replaced by chlorine; a limpid fluid; boils at 167°; has an ethereal odor, and sweet taste.

7. METHYLE SERIES.

A. Formation of wood spirit. As, in the fermentation of sugar, alcohol is formed, so, in the destructive distillation of wood, a substance allied to alcohol is produced; the former is regarded as the hydrated oxide of a base (ethyle), and in the same way the latter is considered as the hydrated oxide of methyle.

B. Methyle, $C_2 H_3$. A hypothetical base, not yet obtained separate.

C. Methyle and oxygen.—Oxide of methyle, $C_2 H_3 O$. This compound (analogous to ether) is formed by distilling wood spirit with sulphuric acid; it exists also naturally as a salicylate in the oil of wintergreen; a colorless gas; sp. gr. 1.59; soluble in water; has an ethereal odor; does not liquefy at 0°. In composition it is isomeric with alcohol, but its equivalent is only half that of alcohol; 2 equivalents of oxide of methyle, or $2\,C_2 H_3 O = 1$ of alcohol, or $C_4 H_6 O_2$.

i. Oxide of methyle and water, hydrated oxide of methyle, pyroxylic spirit, wood spirit, $C_2 H_3 O + H O$. Formed together with acetic acid, in the destructive distillation of wood; a limpid, colorless liquid; sp. gr. 0.800; boils at 150°; readily miscible with water, alcohol, and ether; very inflammable; burning with a pale lambent flame like alcohol, which substance it closely resembles in its solvent powers; it is much used in the arts to dissolve resin, &c., being cheaper than spirit of wine, and also, in some cases, more efficacious. Like oxide of ethyle, the oxide of methyle combines with acids to form salts, or compounds, analogous to the ethers.

ii. Nitrate of oxide of methyle, $C_2 H_3 O + N O_5$. Formed by distilling wood spirit, nitre, and sulphuric acid; a colorless, ethereal liquid; sp. gr.

1.182; boils at 150°; its vapor explodes at temperatures above 250°, being resolved into nitrous acid, water, carbonic acid, and nitrogen.

iii. Sulphite of oxide of methyle, methylosulphurous acid, $C_2 H_3 O + H O + 2 S O_2$.

iv. Sulphate of oxide of methyle, $C_2 H_3 O + S O_3$. Formed by distilling wood spirit and sulphuric acid; a colorless, oily liquid; sp. gr. 1.324; boils at 370°; insoluble in water, but slowly resolved by cold water and rapidly in hot water into wood spirit and a bisulphate of oxide of methyle, or sulphomethylic acid; a white, crystalline solid, $C_2 H_3 O + H O + 2 S O_3$; soluble in water and alcohol, and combining with bases to form salts.

v. Oxalate of oxide of methyle, $C_2 H_3 O + C_2 O_3$. Formed by distilling a mixture of wood spirit, oxalic acid, and sulphuric acid; a white, volatile, crystalline solid; fusible at 124°, and boiling at 322°; soluble in water, which soon resolves it into wood spirit and oxalic acid.

vi. Acetate of oxide of methyle, $C_2 H_3 O + C_4 H_3 O_3$. Exists in the crude wood spirit. Formed by distilling wood spirit, acetic acid, and sulphuric acid; a colorless liquid; easily miscible in water, alcohol, and ether; sp. gr. 0.908; boils at 144°.

vii. Mucate of oxide of methyle, $C_2 H_3 O + C_{12} H_8 O_{14}$. Formed by the action of wood spirit, mucic acid, and sulphuric acid; a fixed, crystalline solid; sp. gr. 1.50; heated it does not fuse, but decomposes at 325°.

D. Methyle and chlorine.—Chloride of methyle, $C_2 H_3 Cl$. Formed by distilling a mixture of chloride of sodium, sulphuric acid, and wood spirit; a colorless, permanently elastic gas; sp. gr. 1.731; and may be collected over water; it has an agreeable ethereal odor, is combustible, and burns with a greenish flame.

E. Methyle and iodine. — Iodide of methyle, $C_2 H_3 I$. Formed by distilling a mixture of phosphorus, iodine, and wood spirit; a colorless liquid; insoluble in water; sp. gr. 2.23; boils at about 110°; methyle also combines with bromine, fluorine, cyanogen, and sulphur, &c.

8. **Formyle Series.**

A. Oxidation of wood spirit. As alcohol by oxidation passes into acetic acid, so wood spirit, by the same process, generates formic acid, $C_2 H_3 O + HO$, passing into $C_2 H O_3$. This is regarded as an oxide of formyle.

B. Formyle, $C_2 H$. A hypothetical base not yet isolated.

C. Peroxide of formyle, formic acid, $C_2 H O_3 + H O$. Obtained by the destructive distillation of ants, or by oxidation of wood spirit by means of finely divided platinum; it has not been obtained separate anhydrous. The hydrated acid is a colorless and slightly fuming liquid; boils at 212°; solidifies in crystals at 32°; its sp. gr. is 1.235; readily dissolves in water; its vapor is combustible: it combines with bases, and forms salts; it also forms a hydrate with 2 equivalents of water, which boils at 222°; does not solidify till 5°, and has a density of 1.1004.

i. Formic acid and ammonia.—Formiate of ammonia, $C_2 H O_3 + N H_3 + H O$. Formed direct; a white, crystalline salt, very soluble in water; fusible at 248°.

ii. Formic acid and oxide of methyle, $C_2 H_3 O + C_2 H O_3$. Obtained by distilling sulphate of methyle and dry formiate of soda; a colorless, transparent liquid; boils at 98°.

D. Formyle and chlorine.—Terchloride of formyle, chloroform, $C_2 H Cl_3$. Obtained by distilling a mixture of chloride of lime with alcohol and water; analogous to formic acid, the whole of the oxygen being replaced by chlorine; a transparent, colorless liquid; sp. gr. 1.480; boils at 140°; slightly soluble in water, easily in ether or alcohol. Its vapor is dense; sp. gr. 4.2. Inspired it produces insensibility, like ether, but even more rapidly. Used as an anæsthetic agent.

E. Teriodide of formyle, iodoform, $C_2 H I_3$. Obtained

by boiling iodine, carbonate of potassa, alcohol, and water; a yellow, crystalline solid; insoluble in water; soluble in alcohol and ether; fuses at 212°, and sublimes.

F. Terbromide of formyle, bromoform, C_2 H Br_3. Obtained like the chloride; a heavy, volatile liquid; sp. gr. 2.10.

G. Tersulphuret of formyle, sulphoform, C_2 H S_3. Obtained by distilling teriodide of formyle and sulphuret of mercury; a dense, oily-looking yellow fluid insoluble in water.

9. AMYLE SERIES.

A. Formation of fusel oil. During the formation of alcohol from potatoes, a quantity of an acrid, oily substance is often produced, which is separated with considerable difficulty, and termed fusel oil; a series of compounds are derived from it, analogous to those of the preceding series; they are traced to a radical, amyle.

B. Amyle, C_{10} H_{11}. A hypothetical radical, not isolated.

C. Amyle and oxygen.—Amyle ether, oxide of amyle, C_{10} H_{11} O. Obtained by heating a mixture of chloride of amyle, potash, and alcohol; a colorless, transparent liquid, boiling at 233°.

i. Oxide of amyle and water.—Hydrated oxide of amyle, oil of potato spirit, fusel oil, C_{10} H_{11} O + HO. Obtained during the rectification of potato spirit; a pale, yellowish, oily fluid; sp. gr. 0.818; boils at 269°; at—4° it solidifies. Its vapor is inflammable, burning with a pale blue flame. It has a powerful and peculiarly suffocating odor.

ii. Salts of oxide of amyle. Amyle, like ethyle and methyle, combines with acids to form ethereal compounds. With sulphuric acid it forms sulphamilic acid, C_{10} H_{11} O + 2 S O_3 + H O, which corresponds to the sulphovinic.

D. Amyle also combines with chlorine, iodine, bromine, &c., to form compounds of $C_{10}\,H_{11}\,Cl$,—$C_{10}\,H_{11}\,Br$, and $C_{10}\,H_{11}\,I$, &c.

E. Oxidation of amyle. As by the oxidation of alcohol acetic acid is produced, by that of wood spirit formic acid, so fusel oil when oxidized is converted into valerianic acid, $C_{10}\,H_{11}\,O + H\,O$ becoming $C_{10}\,H_9\,O_3$. This acid exists naturally in the root of the *Valeriana officinalis*, and may be procured from it by distillation, or obtained artificially from fusel oil and potash, under the continued influence of heat; a transparent, colorless oily liquid; sp. gr. 0.937. it boils at 347°, its vapor burns with a bright but smoky flame; it combines with bases to form valerianates.

10. KAKODYLE SERIES.

A. Analogous to the preceding compounds is a very remarkable set of substances, derived from a compound base, consisting of carbon, hydrogen, and the metal arsenic, which is termed kakodyle.

B. Kakodyle, $C_4\,H_6\,As$. It is obtained by the reaction of pure zinc on the chloride of kakodyle; a limpid colorless liquid, which solidifies in crystals at 20°; sp. gr. 1.48; it inflames spontaneously in the air, oxygen or chlorine; it has a most offensive odor, and, as well as nearly all its compounds, is highly poisonous. Boils at 338°.

C. Kakodyle and oxygen. Oxide of kakodyle, alkarsine, $C_4\,H_6\,As\,O$. Obtained by the distillation of acetate of potassa and arsenious acid; a colorless, limpid liquid; sp. gr. 1.64; freezes at 9°, and boils at 300°; its odor is insupportable, and it is highly poisonous; it inflames spontaneously in the air, chlorine, the vapor of bromine, &c.

D. Kakodylic acid. — Alkargen, $C_4\,H_6\,As\,O_3 + H\,O$. Formed by the action of red oxide of mercury on oxide of kakodyle under water; a brilliant, colorless, crystalline solid; it fuses at 392°, and is decomposed at 446°; it is somewhat deliquescent; easily soluble in water; inodorous; slightly sour, and apparently not at all poisonous; it combines with bases to form salts; amongst the most remarkable of them is the kakodylate of oxide of kakodyle, $C_4 H_6 As O_3 + C_4 H_6 As O$.

E. Kakodyle and chlorine.—Chloride of kakodyle, $C_4 H_6 As Cl$. Formed by the reaction of perchloride of mercury and oxide of kakodyle; a clear, colorless liquid, denser than water; it boils at about 212°; it is very inflammable; has a most offensive and irritating odor, and is eminently poisonous.

F. Kakodyle and iodine.—Iodide of kakodyle, $C_4 H_6 As I$. Obtained by distilling oxide of kakodyle with an aqueous solution of hydriodic acid; a yellow, oily liquid; insoluble in water; boils at about 215°; highly poisonous.

G. Bromide and fluoride of kakodyle resemble the chloride.

H. Kakodyle and sulphur.—Sulphuret of kakodyle, $C_4 H_6 As S$. Obtained by distilling chloride of kakodyle with a solution of the double sulphuret of barium and hydrogen; a colorless transparent liquid; insoluble in water; spontaneously inflammable in the air; and which combines with excess of sulphur to form a persulphuret, $C_4 H_6 As S_3$, which is an acid, forming salts with the other sulphurets.

I. Kakodyle and cyanogen.—Cyanide of kakodyle, $C_4 H_6 As N C_2$. Formed by distilling bi-cyanide of mercury, oxide of kakodyle and water; a brilliant, colorless, crystalline solid; fusible at about 90°; boiling at 284°; has a most offensive odor, and is highly poisonous.

11. BENZOYLE SERIES.

A. Benzoic acid and several allied compounds are traced to a radical termed benzoyle, of which the benzoic acid is an oxide.

B. Benzoyle, $C_{14} H_5$. A hypothetical radical, not isolated.

C. Benzoyle and oxygen. — Benzoic oxide, $C_{14} H_5 O_2$. Obtained by heating the benzoate of copper; a white, crystalline solid; fusible at 160°; insoluble in water, but soluble in ether and alcohol. Benzoic oxide combines with amidogen to form an amide, $C_{14}H_5O_2 + NH_2$, termed benzamide; it is obtained by the action of ammonia on the chloride of oxide of benzoyle; benzamide combines with bromine to form a bromide of benzamide, which may be represented as $C_{14}H_7NO_2 + Br$.

i. Benzoic oxide and hydrogen, $C_{14}H_5O_2 + H$. Exists in distilled bitter almond oil; an aromatic, pungent, colorless liquid; not poisonous when quite free from hydrocyanic compounds; dissolves in water, ether, and alcohol; boils at about 350°; is easily inflammable, and burns with a bright but smoky flame.

D. Benzoyle and oxygen.—Benzoic acid, $C_{14}H_5O_3$. Exists naturally in many resinous plants; obtained particularly from gum benzoin by sublimation, artificially obtained by the oxidation of bitter almond oil; a white, crystalline solid; inodorous when cold; fuses at 250°; boils at 462°; sublimes easily a little above 295°; soluble in water, and crystallizes from its solution as a hydrate with HO; combines with bases to form benzoates; by the action of nitric acid it is converted into nitrobenzoic acid, which may be called benzoic acid, having 1 equivalent of hydrogen replaced by 1 of nitrous acid, $C_{14}H_4O_3 + NO_4 + HO$. Benzoic acid also combines with sulphuric acid to form sulphobenzoic acid, analogous to the sulphovinic; it consists of $C_{14}H_5O_3SO_3 + 2HO$, and readily unites with bases. By the action of chlorine under the influence of solar light three distinct chlorobenzoic acids may be formed; in these respectively, 1, 2, and 3, equivalents of the hydrogen of benzoic acid are replaced by chlorine.

E. Benzone, $C_{13}H_5O$. This compound corresponds to acetone in the acetyle series; it is produced by the destructive distillation of dry benzoate of lime; a colorless, oily fluid; boils at 482°. During its formation there also appears a peculiar hydrocarbon termed benzine or benzole; it consists of $C_{12}H_6$, and is identical with the bicarburet of hydrogen, etherine; a limpid, colorless liquid; sp. gr. 0.850; crystallizing at 32°, and boiling at 186°. It combines with chlorine to form a crystalline, solid compound, consisting of $C_{12}\ H_6\ Cl_6$ (chloride of benzole).

F. Oxide of benzoyle and chlorine.—Chloride of oxide of benzoyle, $C_{14}\ H_5\ O_2\ Cl$. Formed by the action of chlorine on the oil of bitter almonds; a colorless liquid; sp. gr. 1.106; its vapor is inflammable, burning with a green flame.

G. Oxide of benzoyle and iodine.—Iodide of oxide of benzoyle, $C_{14}\ H_5\ O_2\ I$. Obtained by distilling the chloride with iodide of potassium; a colorless, crystalline solid. The oxide of benzoyle also combines with bromine, sulphur, &c.

H. Benzoyle and hydrogen.—Hydruret of benzoyle, picramyle, $C_{14}\ H_6$. Obtained from sulphobenzoyle; a fusible, volatile solid.

12. Salicyle Series.

A. Salicyle, $C_{14}\ H_5\ O_4$. The base of a series of compounds derived from the meadowsweet, willow, poplar, &c.

B. Salicyle and hydrogen.—Salicylous acid, hydruret of Spiroyle, the volatile oil of meadowsweet, $C_{14}\ H_6\ O_4$. Obtained from the meadowsweet (*Spiræa ulmaria*) and from salicine when distilled with sulphuric acid and bichromate of potassa; a colorless, fragrant liquid; slightly soluble in water; congealing at —4°; boiling at 385°; sp. gr. 1.173. By the action of chlorine and bromine, it passes into chlorosalicylous and bromosalicylous acids; and by the action of nitric acid it is converted into nitrosalicylous acid.

C. Salicyle combines with chlorine and bromine.

D. Salicyle and oxygen.—Salicylic acid, $C_{14}H_5O_5$. Formed by the action of fused potash in indigo, by heating salicylous acid with potash, &c.; a white, crystalline solid; fusible at 300°; volatile at higher temperatures; but slightly soluble in water. Chlorine and bromine form peculiar acids with it. Nitric acid converts it into anilic acid. Exists naturally as salicylate of methyle in the essence of *gaultheria procumbens*.

i. Anilic, or nitrosalicylic acid, indigotic acid, $C_{14} H_4 O_5 N O_4 + H O$. Formed by the action of dilute nitric acid on indigo, or of strong nitric acid on salicylic acid, &c.; a white, crystalline solid; soluble in water and alcohol; volatile when heated. Most of its salts are yellow, and explode or deflagrate when heated.

E. Salicine, $C_{26} H_{18} O_{14}$. Obtained from the leaves and young bark of the poplar, willow, &c.; a white, fusible, crystalline solid, which, distilled with acids, yields salicylous acid.

F. Phloridzine, $C_{21} H_{11} O_8 + 4 H O$. A white, crystalline solid, which exists in the roots of the pear, plum, apple, &c.; and which, in several respects, is closely allied to the salicyle series. It fuses at 228°; boils at 380°, when it loses $2 H O$; its sp. gr. is 1.43. By the continued action of hot acids it is converted into phloretine, $C_{30} H_{15} O_{10}$; and by the action of ammonia and oxygen, it passes into phloridzeine, a brilliant, red, uncrystallizable solid, which unites with ammonia and other bases.

13. Albumen, $C_{400} H_{310} N_{50} O_{120} P S_2$. Constitutes the white of egg, exists in the blood of animals, and in plants; a translucent horny solid when dry; softens and gradually dissolves in warm water, particularly if alkali is present; insipid; odorless; as it exists in the animal system dissolved; its solution coagulates by heat.

14. Fibrine, $C_{400} H_{310} N_{50} O_{120} P S$. Constitutes the muscular fibre of animals; exists in their blood, and in the gluten of plants; in the first state it is white; elastic; tasteless; insoluble in water, or when dried; translucent and horny; soluble in alkaline solutions.

15. Caseine, $C_{400} H_{310} N_{50} O_{120} S$. The basis of cheese; exists

in the milk of animals, and in the seeds of leguminous plants; a white, curdy solid; insoluble in water; soluble in dilute acid, and alkaline solutions; its solutions are coagulated in a remarkable manner by certain animal membranes, as in the making of curd and whey; vegetable caseine is termed legumine and emulsine.

16. **Proteine**, $C_{40} H_{31} N_5 O_{12}$. Origin of its name. Obtained from albumen, fibrine, and caseine, by the action of acetic acid, on a warm alkaline solution of either of them; a white, tasteless, insoluble solid.

17. **Gelatine**, $C_{13} H_{10} N_2 O_5$. Obtained from the skin, membranes, tendons, bones, &c., of animals; pure glue; insoluble in cold water; soluble in hot; forms an insoluble compound with tannic acid, hence the art of tanning leather. There is a variety of gelatine obtained from the cartilage of the ribs, called chondrine, $C_{16} H_{13} N_4 O_7$. When gelatine is boiled with alkaline solutions, it is gradually converted into a crystallizable sweet principle, glycocoll, $C_4 H_4 N O_3$.

18. **Diastase.** A modification of vegetable albumen which is formed from it naturally during the germination of seeds, and which possesses the power of causing the conversion of starch into sugar. The change produced in malting grain is caused by the formation of this substance.

19. **Alkaloids or Organic Bases.**

There exist in many plants compounds of carbon, hydrogen, oxygen, and nitrogen, which unite to acids as bases, and form salts. A number of similar compounds may be artificially formed. Some of the more important of these are the following:—

A. Aniline, $C_{12} H_7 N$. Formed by the action of alkalies on indigo, or of H Cl acid and alkali upon the oil of coal tar; a thin oily colorless liquid; boils at 358°; sp. gr. 1.028; soluble in water; forms crystalline compounds with the acids.

B. Brucia, $C_{44} H_{25} O_7 N_2$. Exists in nux vomica; obtained by the action of alcohol and oxalic acid. It crystallizes with 8 equivalents of water; is soluble in water, and combines with acids. The hydrochlorate is a white crystalline salt, soluble in water, and consisting of $C_{44} H_{25} O_7 N_2 + Cl H$. The sulphate forms very soluble efflorescent crystals, consisting of $C_{44} H_{25} O_7 N_2 + S O_3 + 5 H O$.

C. Benzoline, $C_{42} H_{18} N_2$. Formed by the action of alkalies on bitter almond oil; a white solid; insoluble in water, readily soluble in spirit; fusible below 212°; forms definite, crystalline salts with the acids.

D. Codeia, $C_{35} H_{20} O_5 N$. Exists in opium; a colorless solid; fusible at 300°; burns with a smoky flame; slightly soluble in water; crystallizes with 2 equivalents of water; forms salts.

E. Caffeine, theine, $C_8 H_5 O_2 N_2$. Exists in tea and coffee; a white crystalline solid; soluble in warm water, and forms salts.

F. Chinchonia, $C_{20} H_{12} O N$. Exists in pale Peruvian bark; a white, semi-transparent, crystalline solid; but little soluble in water; unites with acids; its salts for the most part are remarkably bitter. The muriate may be formed direct, and consists of $C_{20} H_{12} O N + Cl H$. It also forms a subsalt of $2 C_{20} H_{12} O N + Cl H$, which crystallizes in acicular prisms.

G. Conia, $C_{16} H_{16} N$. Exists in hemlock; a colorless, volatile oil, of sp. gr. 0.89; highly poisonous, and forming difficultly crystallizable salts.

H. Furfurine, $C_{30} H_{12} O_6 N_2$. Formed by the action of ammonia on an oil obtained from bran, distilled with dilute sulphuric acid; a white, crystalline solid; sparingly soluble in water, easily soluble in alcohol and ether; burning when strongly heated, and uniting to acids to form crystallizable salts.

I. Meconine, $C_{10} H_5 O_4$. Exists in opium; a white, inodorous solid; soluble in water, alcohol, and ether; fusible at 195°; distils at 350°; unites with acids to form salts.

K. Menispermia, $C_{18} H_{12} O_2 N$. Exists in cocculus indicus; a white, crystalline solid; fusible at 250°; insoluble in water, soluble in alcohol and ether; forms crystalline salts with the acids. There is also an isomeric compound called paramenispermia, fusible at 482°, but not salifiable.

L. Morphia, $C_{35} H_{20} O_6 N$. Exists in opium; a white solid; nearly insoluble in water, but deposited from its solutions in crystals containing 2 equivalents of water. It readily forms crystalline salts with the acids. The sulphate contains $C_{35} H_{20} O_6 N + S O_3 + 6 H O$; it also forms a bisulphate. The hydrochlorate may be formed direct; it is an anhydrous salt, containing $C_{35} H_{20} O_6 N + Cl H$.

M. Narceia, $C_{28} H_{20} O_{12} N$. Exists in opium; a white, crystalline solid; it dissolves in water, and in acids, without neutralizing them.

N. Nicotine, $C_{10} H_8 N$. Exists in tobacco; a colorless oily liquid; sp. gr. 1.048; volatile below 475°; inflammable; highly poisonous; combines with acids, and forms difficultly crystallizable salts.

O. Picrotoxia, $C_{12} H_7 O_5$. Exists in cocculus indicus; a colorless, crystalline solid; soluble in water and alcohol; intensely bitter; forms crystallizable compounds with acids.

P. Piperine, $C_{34} H_{19} O_6 N$. Exists in black pepper; a pale, yellow, crystalline solid; inodorous and tasteless; fuses about 215°; forms salts.

Q. Paramorphia, thebaia, $C_{25} H_{14} O_3 N$. Exists in opium; a white solid; crystallizing with two equivalents of water, and salifiable.

R. Pseudomorphia, $C_{27} H_{18} O_{14} N$. Exists in opium; forms salts.

S. Quinia, quinine, $C_{20} H_{12} O_2 N$. Exists in yellow bark; a white, fusible solid; slightly soluble in water, with which it forms a crystalline hydrate; it forms crystalline salts with acids. The sulphate, $C_{20} H_{12} O_2 N + S O_3$, may be obtained anhydrous, or crystallized with 8 equivalents of water; the disulphate,

$2\ C_{20}\ H_{12}\ O_2\ N + S\ O_3$ is anhydrous, or crystallized with 2, 4, or 8 equivalents of water; there is a basic and a neutral hydrochlorate; the salts of this alkaloid are much used in medicine. There is also a substance called quinoidine, or amorphous quinine, stated to be identical in composition with quinine; it is uncrystallizable.

T. Quinoline, $C_{19}\ H_8\ N$. (Leucolin?) Formed by the action of caustic potash on quinia, strychnia, or cinchonia; an oily liquid; sp. gr. 1.084; forms salts.

U. Sinnamine, $C_8\ H_6\ N_2 + H\ O$. Formed by the action of bases on the volatile oil of black mustard; a crystalline, colorless solid; forms salts, of which the oxalate alone is crystallizable.

V. Solania, $C_{84}\ H_{68}\ O_{28}\ N$. Exists in the potato tribe; a white, crystalline solid; highly poisonous; salifiable, and most of its salts uncrystallizable.

W. Strychnia, $C_{44}\ H_{24}\ O_4\ N_2$. Exists in nux vomica; a crystalline, colorless solid; intensely bitter; highly poisonous; forms soluble, crystallizable salts.

X. Theobromine, $C_{14}\ H_8\ N_4\ O_4$. Exists in cocoa nuts; a white, crystalline solid; forms salts.

Y. Toluidine, $C_{14}\ H_9\ N$. Obtained from the oil of tolu-balsam, by the action of nitric acid, &c.; a colorless, crystalline solid; fuses at 104°; boils at 388°. It forms crystalline salts with acids.

Z. Urea, $C_2\ H_4\ N_2\ O$. Identical with cyanate of ammonia; exists naturally in the urine of animals; a colorless, transparent, crystalline solid; soluble in water and alcohol; sp. gr. 1.35; fuses at 250°. It forms crystallizable salts.

AA. Veratria, $C_{34}\ H_{22}\ O_6\ N$. Exists in hellebore; a colorless solid; fuses at 122°; forms salts with acids, which crystallize with difficulty.

20. Compound Acids of Organic Origin.

A. Aconitic acid, $C_4\ H\ O_3 + H\ O$. Exists in aconite; artificially obtained by action of heat on citric acid; soluble in water; fuses at 285°; cannot be obtained anhydrous and independent of a base.

B. Amygdalic acid, $C_{40} H_{26} O_{24} + HO$. Formed by the action of baryta on amygdalene; a principle obtained from the bitter almond.

C. Aspartic acid, $C_8 H_5 O_6 N + H O$. Formed by the action of bases on asparagine; tasteless and inodorous; forms aspartates.

D. Butyric acid, $C_8 H_7 O_3 + H O$. Exists in butter; a colorless, oily fluid; sp. gr. 0.963; boils at 327°, and does not solidify at 0°. Its salts are, for the most part, crystallizable.

E. Carbazotic acid, picric acid, $C_{12} H_2 N_3 O_{13} + H O$. Formed by the action of nitric acid on indigo; a bright yellow, crystalline solid; fusible and volatile when heated; combustible; forms salts, most of which detonate, when heated, with considerable violence.

F. Citric acid, $C_{12} H_5 O_{11}$. Exists in lemons, limes, currants, and other fruits; a white solid; very soluble in water, and crystallizing from its solution with 3, 4, or 5 equivalents of water. Like phosphoric acid, it forms three classes of salts, containing 3 equivalents of base, 2 equivalents and 1 of water, and 1 of base and 2 of water. Strongly heated, citric acid is converted into the aconitic; during this action, two other acids are formed:—itaconic, $C_5 H_2 O_3 H O$; a transparent, crystalline solid; fusible at 320°, which may be rendered anhydrous:—and citraconic acid, which is isomeric with the preceding; a limpid, oily liquid.

G. Choleic acid, $C_{44} H_{40} N S O_9$. Exists combined with soda as a choleate of soda in animal bile, a secretion of the liver; an amorphous, brittle, yellow solid; soluble in water and alcohol, and having an intensely bitter taste.

H. Gallic acid, $C_7 H_2 O_4 + H O$. Exists in oak galls, &c.; a white, crystalline solid; soluble in water. The acid is bibasic, its salts containing 2 equivalents of base to 1 of acid. Associated with this acid is another, closely allied to it, and probably isomeric, called ellagic acid,

$C_7 H_2 O_4$; it is insoluble in water. When heated, gallic acid is converted into pyrogallic acid, $C_6 H_3 O_3$; a white, crystalline solid; fusible at 240°, and volatile at 410°.

I. Hippuric acid, urobenzoic acid, $C_{18} H_8 O_5 N + H O$. Exists in the urine of many animals; a colorless, fusible, crystalline solid. It yields benzoic acid, when distilled.

K. Kinic acid, quinic acid, cinchonic acid, $C_7 H_5 O_5 + H O$. Exists in the cinchona bark; a colorless, crystalline solid; soluble in water; fuses at 310°; the salts are mostly crystallizable, and soluble.

L. Malic acid, sorbic acid, $C_8 H_4 O_8 + 2 H O$. Exists in apples, the berries of the mountain ash, rhubarb stalks, &c.; a very deliquescent, semi-crystallized solid; soluble in water and alcohol. Malic acid is bibasic; its salts contain 2 equivalents of base to 1 of acid.

M. Margaric acid, $C_{34} H_{33} O_3 + H O$. Obtained from various solid oils by the action of bases; or from stearic acid by the action of nitric acid; a white, inodorous, crystalline, solid fat; fuses at 140°; insoluble in water; soluble in boiling ether and alcohol; when heated, it distils over unchanged, a portion of margarone, $C_{33} H_{33} O$, being formed. By the action of sulphuric acid, two new modifications of margaric acid are formed, containing 4 and 5 equivalents of oxygen.

N. Meconic acid, $C_{14} H O_{11} + 3 H O$. Exists in opium; a transparent, colorless, crystalline solid; soluble in alcohol and water; it is not decomposed by a heat of 240°; but its aqueous solution is decomposed by boiling. It forms three classes of meconates:—those with 3 equivalents of base to 1 of acid; those with 2 of base and 1 of water; and those with 1 of base and 2 of water.

O. Oleic acid, $C_{36} H_{33} O_3 + H O$. Obtained from various fat oils; a colorless, oily liquid; not miscible with water, but easily soluble in alcohol. It is decomposed by heat, sebacic acid being formed. By oxidizing agents it is transformed into a series of new acids.

P. Pinic acid, $C_{40} H_{30} O_4$. Constitutes the greater part of common resin or colophony; a colorless, transparent, uncrystallizable solid; insoluble in water, but soluble in alcohol, ether, &c. There is also an isomeric form, termed silvic acid; it is a crystalline solid, which fuses at 285°; its salts are crystallizable, and soluble in water. When the alcoholic solution of this acid is exposed to the air, it becomes converted into an uncrystallizable, oxysilvic acid, $C_{40} H_{29} O_7$.

Q. Racemic acid, $C_4 H_2 O_5$. Exists in grapes; a white solid; readily soluble in water, and crystallizing with 2 equivalents of water. Analogous to tartaric acid.

R. Sebacic acid, $C_{10} H_8 O_3 + H O$. Obtained by the distillation of tallow; a white, volatile, crystalline solid; soluble in water and alcohol. Fuses at 260°, and forms salts with the bases.

S. Suberic acid, $C_8 H_6 O_3 + H O$. Formed by the action of nitric acid on cork, tallow, &c.; a white solid; fusible at about 250°: volatile at a higher temperature; soluble in alcohol and water.

T. Succinic acid, $C_4 H_2 O_3 + H O$. Obtained by the distillation of amber, or by the action of hot, strong nitric acid on stearic acid. Anhydrous succinic acid is a white solid, fusible at 295°, and boiling at 480°. The hydrate is crystalline, and of sp. gr. 1.55; fuses at 320°, and boils at 470°; it is soluble in water, ether, and alcohol.

U. Tannic acid, tannin, $C_{18} H_8 O_{12}$. Exists in oak galls and bark, and most astringent vegetables; extracted from powdered galls by ether; a white, inodorous solid; very soluble in water; having a powerfully astringent taste; precipitates solutions of animal glue; when acted on by the alkalies, it yields oxytannic acid, $C_{15} H_8 O_{14}$.

V. Tartaric acid, $C_8 H_4 O_{10} + 2 H O$. Exists in many fruits; obtained from the tartar of wine, which is a bitartrate of potash; a semi-transparent, crystallized solid; sp. gr. 1.70; very soluble in water; used in calico printing; when heated up to 392°, it loses one-fourth of its water, and acquires new properties; it is then called tartralic acid. If tartralic acid is kept for some time at a temperature of 356°, it loses more water, becoming a monohydrate, when it is called tartrelic acid. When tartar is distilled, it yields pyrotartaric acid, which consists of $C_5\ II_3\ O_3 + H\ O$. During the distillation of tartaric and also racemic acid, a quantity of pyruvic acid is formed, which consists of $C_6\ H_3\ O_5 + H\ O$.

W. Uric or lithic acid, $C_{10}\ H_4\ O_6\ N_4$. Exists in the urine of various animals, particularly in that of serpents; a white, crystalline solid; tasteless, inodorous, and nearly insoluble in cold water; its salts are chiefly interesting in connection with the study of animal chemistry. When uric acid is boiled with peroxide of lead, and water, it is converted into Allantoin, $C_4\ H_3\ N_2\ O_3$, a substance which occurs naturally in the allantoic liquor of the fœtal calf; it is a brilliant, crystalline, colorless solid, possessing neither acid nor basic properties. Cold nitric acid converts uric acid into Alloxan, $C_8\ H_4\ N_2\ O_{10}$; it is a brilliant, colorless, crystalline solid, which by the action of bases is converted into Alloxanic acid, $C_4\ H\ N\ O_4 + H\ O$; and by the action of heat there is also formed Mesoxalic acid, $C_3\ O_4 + 2\ H\ O$, and Mykomelinic acid, $C_8\ H_5\ N_4\ O_5$. Hot, strong nitric acid converts uric acid and also Alloxan into Parabanic acid, $C_6\ N_2\ O_4 + 2\ H\ O$. When the salts of this acid are heated, the acid is changed into the Oxaluric, $C_6\ H_3\ N_2\ O_7 + H\ O$. Hot nitric acid, when dilute, converts uric acid into Alloxantine, $C_8\ H_5\ N_2\ O_{10}$, a transparent, colorless, crystalline solid. When ammonia is added to a hot mixed solution of alloxantine and alloxan, murexide or purpurate of ammonia is formed, $C_{12}\ H_6\ N_5\ O_8$; a brilliant, crystalline solid; deep purple red by transmitted light, but a brilliant green by reflected light. By the action of strong acids, murexide is decomposed, and purpuric acid obtained; a yellow crystalline solid.

21. INDIFFERENT PRINCIPLES.

Besides the compounds derived from existing or hypothetical bases, the alkaloid series, and the compound organic acids, there are a great number of substances allied more or less to them, but not directly classified, which may be conveniently considered in this place.

OILS.

These are divided into fixed or fat, and volatile. The fixed oils are further divided into drying or non-drying oils. Fat oils consist of two or three distinct oily substances, termed oleine, stearine, and margarine. These are compounds of three peculiar acids with glycerine. Some contain volatile acids, to which their peculiar odor is due.

A. Glycerine, $C_6 H_7 O_5 + HO$. Exists in fixed oil; obtained during the process of saponification; a thick, viscid liquid; sp. gr. 1.28; easily soluble in water; intensely sweet; uncrystallizable; does not undergo the alcoholic fermentation; yields acrolein by heat.

B. Oleine, oleate of glycerine, $C_{44} H_{40} O_4 + C_6 H_7 O_5$. Constitutes the most fluid part of oil; sp. gr. about 0.9; colorless, inodorous, tasteless, oily fluid; does not solidify by cold.

C. Stearine, stearate of glycerine, $C_{68} H_{66} O_5 + C_6 H_7 O_5$. Forms the more solid part of oil; best obtained from the solid fats, as tallow; a translucent, brittle, semi-crystalline, fatty solid; fusible at about 145°; insoluble in water and cold alcohol; used for candles.

D. Margarine, margarate of glycerine, $C_{84} H_{33} O_3 + C_6 H_7 O_5$. Exists in many oils with stearine; a white, crystalline solid; fusible at 120°. These three compounds, when saponified, or boiled with a base, are decomposed, new salts or soaps of the fatty acids being formed whilst the glycerine is set free.

E. Non-drying fat oils, liquid at common temperatures. The chief of these are, 1. Almond (*Amygdalus communis*); sp. gr. 0.918; congeals at 30°. 2. Beech mast (*Fagus sylvaticus*); sp. gr. 0.922; congeals at 29°. 3. Colza (*Brassica campestris*); sp. gr. 0.912; congeals at 21°. 4. Ground nut (*Arachis hypogæa*); sp. gr. 0.916; congeals at 27°. 5. Hazel nut (*Corylus avellana*); sp. gr. 0.924; congeals at 30°. 6. Olive (*Olea europæa*); sp. gr. 0.910; congeals at 36°. 7. Black mustard (*Sinapis nigra*); sp. gr. 0·917; congeals at 30°. 8. White mustard (*Sinapis alba*); sp. gr. 0.915; congeals at 30°. 9. Rape (*Brassica rapa*); sp. gr. 0.916; congeals at 28°; and (*Brassica napus*); sp. gr. 0.912; congeals at 26°. 10. Till (*Sesamum orientale*); sp. gr. 0.917. 11. Whale or spermaceti oil; sp. gr. 0.927.

F. Non-drying oils, solid at common temperatures. The chief are, 1. Bassia butter (*Bassia butyracea*); sp. gr. 0.942; melts at 120°. 2. Cocoa nut (*Cocos nucifera*); sp. gr. 0.918; fuses at 74°. 3. Chocolate (*Theobroma Cacao*); fuses at 120°. 4. Cows' butter; sp. gr. 0.908; melts about 90°. 5. Hog's lard; sp. gr. 0.930; fuses at 99°. 6. Ox fat; fuses at 100°. 7. Palm oil (*Elais guineensis*); fuses at 85°. 8. Piney tallow (*Vateria indica*); sp. gr. 0.926; melts at 98°. 9. Spermaceti; sp. gr. 0.943; fuses at 112°. 10. Tallow, sheeps' fat; sp. gr. 0.913; melts at 100° to 106.

G. Drying oils. 1. Castor (*Ricinus communis*); sp. gr. 0.960°; solidifies at 0°. 2. Hemp (*Cannabis sativa*); sp. gr. 0.927; solidifies at —17°. 3. Linseed (*Linum usitatissimum*); sp. gr. 0.939; solidifies at —17°. 4. Poppy (*Papaver somniferum*); sp. gr. 0.924; solidifies at 0°. 5. Walnut (*Juglans regia*); sp. gr. 0.928; solidifies at 15°.

H. Uses of oils in the arts. For burning in lamps, &c.; manufacture of candles; stearine candles; manufacture of soaps; manufacture of "drying oil" for printers' ink and varnishes.

I. Wax. Bees' wax; sp. gr. 0.960; fuses at 150°. Myrtle wax; sp. gr. 1.015; fuses at 102°.

J. Volatile oils. These may be divided into those which are pure hydrocarbons, and those which contain other elements in addition; the former have been noticed amongst the hydrocarbons. The more important amongst the latter, consisting of carbon, hydrogen, and oxygen, are, 1. Oil of cloves (*Caryophyllus aromaticus*); sp. gr. 1.050—1.060. 2. Cumin (*Cuminum Cyminum*); sp. gr. 0.860; boils at 540°. 3. Anise (*Pimpinella Anisum*); sp. gr. 0.986; fuses at about 65°, and boils at 430°. 4. Peppermint (*Mentha piperita*); sp. gr. 0.907. 5. Valerian (*Valeriana officinalis*). 6. Caraway (*Carum Carui*). 7. Lavender (*Lavandula vera*). Amongst the volatile oils containing sulphur, are the following: 1. Volatile oil of black mustard, $C_8 H_5 N S_2$; sp. gr. 1.015. Closely allied to this, are the oils of horseradish and scurvy grass. 2. Oil of garlic (*Allium sativum*), $C_6 H_5 S$; boils at about 300°. 3. Oil of assafœtida; consists of carbon, hydrogen, oxygen, and sulphur; sp. gr. 0.942; boils at 320°—370°.

Resins.

These are compounds of carbon, hydrogen, oxygen. As they occur in nature, they are, for the most part, mixtures of various hydrocarbons, together with different oxidized hydrocarbons. The more important are—

A. Solid resins. 1. Colophony, or common resin; sp. gr. 1.080; fuses at 160°; very inflammable; a mixture of pinic and silvic acids. 2. Elemi; sp. gr. 1.018; a mixture of two resins; one $C_{40} H_{32} O_4$; the other, $C_{40} H_{33} O$. 3. Anime (*Hymenæa Courbaril*); sp. gr. 1.028. 4. Sandarach (*Thuia articulata*); sp. gr. 1.060; fuses at 308°. 5. Mastic (*Pistacia Lentiscus*); sp. gr. 1.074; fuses at 250°. 6. Benzoin (*Styrax Benzoë*); sp. gr. 1.063—1.092. 7. Guaiacum (*Guaiacum officinale*); sp. gr. 1.20—1.23; fuses at 570°. 8. Copal (*Hymenæa verrucosa*); sp. gr.

1.045—1.139. 9. Dammar (*Pinus Dammara*); sp. gr. 1.056 to 1.097; fuses at 212°. 10. Lac, produced on the young branches of trees by the *Cocus Ficus;* a mixture of several resins. 11. Amber imbedded in wood coal, &c.; a fossil resin; sp. gr. 1.065.

B. Semifluid resins or balsams, mixtures of resins with volatile oils. 1. Turpentine (*Pinus Abies*, &c.); consists of oil of turpentine, colophony, and other resins. 2. Copaiva (*Copaifera officinalis*, &c.); sp. gr. 0.950; boils at 504°. 3. Balsam of Peru (*Myroxylon Peruiferum*); sp. gr. 1.150; boils at 550°. 4. Styrax (*Liquidamber Styraciflua*). 5. Balm of Mecca (*Amyris Opobalsamum*); sp. gr. 0.950.

C. Gum resins, mixtures of various resins with volatile oils and gum. 1. Ammoniac (*Dorema Ammoniacum*); sp. gr. 1.207. 2. Galbanum (*Galbanum officinale*); sp. gr. 1.212. 3. Assafœtida (*Ferula Assafœtida*); sp. gr. 1.327. 4. Sagapenum. 5. Olibanum. 6. Myrrh; sp. gr. 1.360. 7. Euphorbium. 8. Bdellium. 9. Aloes. 10. Scammony. 11. Gamboge. 12. Opium.

Coloring Matters.

These are very various in their nature and properties. The following are some of the more important: Those which are used in dyeing and calico printing are either substantive or adjective; that is to say, they are either absolute dyes, or are used with the assistance of some other substance which fixes the color in the fibre to be dyed; these substances are termed mordaunts. Alumina, the oxides of iron and tin, are the commonest mordaunts.

A. Blue coloring matters.

i. Indigo, $C_{16}H_5O_2N$. Exists in various plants; obtained chiefly from the *Indigofera*. As met with in commerce, it is impure; it may be purified by sublimation; an intensely deep blue, crystalline, tasteless, inodorous solid; sp. gr.

1.350; volatilizes about 550°, as a deep red violet vapor; burns with a bright flame; insoluble in water, ether, and alcohol. By the action of deoxidizing agents indigo is converted into Indigogene, $C_{16} H_6 O_2 N$; a white solid; tasteless; inodorous; not volatile; readily soluble in alkaline solutions, which, when exposed to the air, absorb oxygen, and the white soluble indigogene reverts to blue, insoluble indigo. This fact is of the first importance in dyeing. Indigo blue is also soluble in strong sulphuric acid, forming a blue compound, hyposulphindegotic acid, $C_{16} H_4 O_2 N + S_2 O_5 + H O$. By the action of nitric acid, indigo is converted into Isatine, $C_{16} H_5 O_2 N + O_2$; a brilliant, red-brown, crystalline solid, which combines with chlorine, iodine, &c., and, when heated with bases, forms isatinic acid, $C_{16} H_6 O_5 N + H O$; nitric acid likewise forms indigotic, or nitrosalicylic acid, and nitropicric or carbazotic acid.

ii. Litmus, archil, cudbear. Derived from various lichens, particularly *Roccella tinctoria*. The principal substances derived from litmus are Lecanorine, $C_{18} H_8 O_8 + H O$; a white, tasteless, inodorous, crystalline solid; insoluble in water, but possessing acid powers. By the action of heat it is converted into Orcine, $C_{16} H_8 O_4 + H O$; a yellowish-white, crystalline solid; soluble in water; volatile at 550°. Ammonia converts orcine into orceic acid, $C_{16} H_8 O_8 N + H O$, which forms deep red or purple salts. It is to the formation of this substance that litmus owes its value in dyeing. Litmus is valuable to the chemist as a test for acids and alkalies; the former render it red, the latter green; test-papers are made by steeping bibulous paper in a strong solution of litmus; the color is very fugitive, being destroyed by exposure to light.

B. Red coloring matters.

i. Madder; the root of the *Rubia tinctorium.* The valuable part of this is the Alizarine or garancine, $C_{37} H_{12} O_{10}$; an orange-colored, crystalline solid; it is an excellent dye. In India, munjeet, the root of *Rubia Munjista,* is much employed as a red dye.

ii. Brazil wood; *Cæsalpinia Sappan,* &c. It contains an orange, crystalline solid, termed Brasileine, $C_{36} H_{14} O_{14}$.

iii. Logwood; *Hæmatoxylon campechianum.* Contains Hæmatoxyline, $C_{40} H_{17} O_{15}$; a brown yellow solid, which crystallizes with 8 equivalents of water, and becomes red under the influence of ammonia, being converted into Hæmateine, $C_{40} H_{14} O_{15} + HO$; a dark green solid when dry, but rendered red by acids, and violet-purple by ammonia.

iv. Cochineal; the female of a species of coccus, which feeds on the nepal (*Opuntia Coccinillifera*). It contains a deep crimson red solid, soluble in water, and having a strong affinity for alumina, termed Carminic acid, $C_{28} H_{14} O_{16}$. Cochineal is chiefly used in the manufacture of carmine and the best red ink.

v. Safflower; the dried petals of the *Carthamus tinctoria.* It derives its color from Carthamic acid, $C_{29} H_9 O_7$; a dark red powder; insoluble in water, but soluble in alkaline solutions. It has a very strong affinity for the fibre of silk and cotton.

vi. Alkanet; the root of the *Anchusa tinctoria.* It contains anchusic acid, $C_{17} H_{10} O_4$; a dark red solid; insoluble in water.

vii. Lac; the coloring matter of the resin lac prepared in India from stick lac.

viii. Annotta; obtained from the seeds of the *Bixa Orellana.* An orange powder, insoluble in water. Its coloring matter, Orelleine, is red, becomes deep blue by the action of sulphuric acid, and dissolves in alkaline solutions. Annotta is chiefly used for coloring cheese.

C. Yellow colors.

i. Quercitron; the bark of the *Quercus nigra*. Contains a yellow, crystalline solid, Quercitreine, $C_{35} H_{10} O_{18} + 2 H O$.

ii. Fustic; the wood of the *Morus tinctoria*. Contains a yellow, crystalline substance, Morine.

iii. Weld; dried plants of *Reseda luteola*. Contains a yellow, crystalline substance, Luteoleine.

iv. Turmeric; the root of *Curcuma longa*. Contains a yellow coloring matter, Curcumine. It is valuable to the chemist as a reagent or test; alkalies and alkaline solutions turn it deep brown; muriatic acid and boracic acid also produce this effect; all other acids restore the yellow color of browned turmeric.

v. Sumach; the young branches of *Rhus coriaria*. Contains a bitter yellow principle, called Fustine.

vi. Persian berries; the dried fruit of the *Rhamnus infectorius*. Contains a brilliant but fugitive yellow coloring principle.

vii. Saffron; the dried stigmas of the flowers of the *Crocus sativus*. Contains a brilliant yellow coloring matter, termed Polychroite.

viii. Barberry; the root and wood of the barberry (*Berberis vulgaris* and *tinctoria*). Contains a peculiar yellow coloring matter, Berberine, $C_{33} H_{18} O_{12} N$; the root is employed in dyeing leather yellow.

D. Green colors.

i. Chlorophylline; the coloring matter of the leaves of plants. A dark green solid; fusible at 390°: insoluble in water; and consisting of $C_{18} H_9 O_8 N$.

ii. Erythrophylline; a modification of chlorophylline, which is found in those leaves which turn red in the autumn. It is turned red by acids, and restored to its original green by alkalies.

22. CHEMISTRY OF VEGETABLE LIFE.

A. The Food of Plants.—The Soil. A mixture of silica, alumina, lime, magnesia, organic matters, oxide of iron, with small quantities of phosphates, carbonates, sulphates, nitrates, and muriates of the alkalies. Its mechanical structure varies greatly, and influences its nature almost as much as its chemical composition.—The air. It contains carbonic acid, salts of ammonia, moisture: these things are not only absorbed direct by plants, but, by their action on the soil, assist in rendering its constituents available as food for growing plants.

B. Growth of Plants.—Germination. A vital point or embryo, surrounded by appropriate food, and placed under suitable conditions, grows. The conditions necessary for germination are, a temperature above 40°, moisture, and oxygen to cause those chemical changes which are necessary. Light is prejudicial, Diastase, or a similar substance, is formed; the starch, &c., becomes soluble; the embryo grows, and a young plant is formed. When it has roots and leaves, it absorbs carbonic acid, ammonia, and water from the soil, together with various inorganic salts.

C. Under the influence of light organic matter is formed, woody fibre and starch, &c., are deposited, and the plant grows and increases, giving off the surplus oxygen from its leaves. In the formation and ripening of seed a slight difference is observed: these parts generally do not evolve oxygen. When plants have completed their growth, and formed seed, they die, ceasing to decompose carbonic acid and water, &c.; the oxygen of the air reacts on the organic matter they have formed; it slowly decomposes, giving out carbonic acid and water, and undergoing eremacausis or decay. Besides the gaseous food of plants, particular earthy and saline matters are necessary to plants; hence they exhaust the soil, when constantly cultivated in the same place. To remedy this, rotations of crops, and various manures are employed, or the soil is left more or less fallow, that, by the action of the air on the stony particles it contains, it may acquire a fresh supply of the salts, &c., the plants have removed.

23. CHEMISTRY OF ANIMAL LIFE.

A. Composition of animals.—Hard solids. Bone; phosphate and carbonate of lime, with gelatine and oil.—Cartilage; gelatine, with a little phosphate of lime.—Soft solids. Muscle, chiefly of fibrine.—Skin, membrane, chiefly gelatine.—Nervous matter, chiefly albumen and oily matter.—Fat, various oily acids.—Liquid component. Blood; water, albumen, saline matter, a trace of fat, and a peculiar, red, globular matter, consisting chiefly of albumen, fibrine, and a red coloring substance, containing oxide of iron. The color of the blood varies; in the arteries it is of a brilliant red; in the venous system it is dark purple, recovering its florid red color again when it passes through the vessels of the lungs.

B. Animal secretions.—Milk; water, fat, caseine, sugar, albumen, saline matters.—Bile; water, choleate of soda, resin, mucus, saline matters.—Saliva; water, traces of mucus, organic matters, salts, &c.—Urine; water, urea, uric acid, albumen, salts of phosphoric, carbonic, and sulphuric acids, combined with lime, ammonia, potash, soda, and magnesia.

C. Chemical processes of life.—Respiration; action of the lungs, to remove carbon from the blood in the form of carbonic acid, heat being evolved.—Mastication; the comminution of food, and mixture of it with air and saliva.—Digestion; a species of fermentation under the influence of a peculiar substance, to which the name of Pepsin has been given, and an acid, probably muriatic; the food is thus converted into chyme; it then becomes mixed with bile, the secretion of the liver; separated into chyle, which is absorbed by the lacteals, and the residue, which is passed from the system as excrementitious; eventually the chyle enters into the venous system, and forms part of the blood. The part of food thus absorbed is divided into that which goes to increase or supply waste in the system; that which is at once burnt or consumed in the production of animal heat by respiration; and that which is stored up as fat.

PART III.

CHEMISTRY OF THE METALLIC ELEMENTS.

SECT. I.—METALS OF THE ALKALIES AND ALKALINE EARTHS.

1. POTASSIUM. Kalium. Symb. K. Eq. 40.

A. Exists as an oxide in many rocks and stones; occurs also in most plants, and is obtained by lixiviating their ashes. The oxide is decomposed by iron or charcoal at a high temperature, and the metal obtained.

B. Properties.—A white, silvery-looking metal, very oxidizable, and hence must be kept out of contact of air or substances containing oxygen; best preserved in naphtha (C H); brittle and crystalline at 32°; soft like wax at common temperatures; fuses at 150°, and boils at a bright red heat, forming a green vapor; sp. gr. 0.865; burns, with a pinkish white flame, when heated in the air, or when thrown on the surface of water or ice, passing into the state of oxide.

C. Potassium and oxygen.—Oxide of potassium, potassa, $K\,O = 40 + 8 = 48$. May be formed direct, or obtained by strongly heating the hydrate of potash with potassium; a hard, brittle, gray solid; sp. gr. 2.65; fusible at a bright red heat; strongly alkaline and caustic; with water it forms a hydrate, $K\,O + H\,O$, common potash. This is usually obtained by decomposing the carbonate of potash by lime. A white, semi-crystalline solid; sp. gr. 1.7; fusible at about

700°; volatile at a bright red heat; very deliquescent, and soluble in water and alcohol. Its solution is caustic, highly alkaline, and reddens vegetable yellows strongly. Potash is a powerful base, and readily forms salts with acids. There is also a suboxide, K_2O; and a peroxide of potassium; formed when the metal is burnt in air or oxygen; it is a yellow, fusible solid; consists of KO_3; is resolved into KO and O_2 by heat, or by solution in water, and is not salifiable.

i. Potash and hypochlorous acid.—Hypochlorite of potassa, KO, ClO (Eau de Javelle). A white, soluble salt, formed direct, and possessing powerful bleaching properties; very easily decomposed.

ii. Potash and chlorous acid.—Chlorite of potassa, KO, ClO_3. A white, crystalline, soluble salt; formed direct.

iii. Potash and hypochloric acid.—Hypochlorate of potassa, KO, ClO_4. A white, deliquescent, crystalline salt; formed direct.

iv. Potash and chloric acid.—Chlorate of potassa, KO, ClO_5. A white, crystalline salt; formed by the action of chlorine on a cold solution of potash or its carbonate; fuses at 500°; by a higher temperature it is decomposed, giving off 40 per cent. of pure oxygen; an action which is remarkably assisted by the *presence* of peroxide of manganese; sp. gr. 1.989; soluble in water; forms fulminating mixtures with combustibles; used in the manufacture of detonating powders, fireworks, matches, &c.

v. Potash and perchloric acid.—Perchlorate of potassa, KO, ClO_7. Formed by the action of sulphuric acid on the chlorate, or its partial decomposition by heat; a white, fusible, crystalline salt, resembling the chlorate; when heated, it yields 45 per cent. of pure oxygen.

vi. Potash and iodous acid.—Iodite of potassa, KO, IO. Formed by the action of a dilute solution of potash on iodine; a transparent, white, crystalline solid.

vii. Potash and iodic acid.—Iodate of potassa, KO, IO_5. Formed direct, and by action of iodine on solution of potassa; a white, crystalline solid, decomposed by a red heat with the evolution of oxygen; insoluble in alcohol; soluble in water. There is also a biniodate and teriodate, containing 2 and 3 equivalents of acid.

viii. Potash and periodic acid. Periodate of potassa, KO, IO_7. A white salt, like the perchlorate.

ix. Potash and bromic acid.—Bromate of potash, $KO, Br O_5$. A white, crystalline solid; formed like the iodate.

x. Potash and hydrochloric acid form chloride of potassium and water, $KO + ClH = KCl + HO$.

xi. Potash and hydriodic acid form iodide of potassium, &c.

xii. Potash and hydrobromic acid form bromide of potassium, &c.

xiii. Potash and hydrofluoric acid form fluoride of potassium, &c.

xiv. Potash and hyponitrous acid. — Hyponitrite of potassa, KO, NO_3. Formed when nitrate of potash is strongly heated; a colorless, slightly deliquescent, crystalline salt.

xv. Potash and nitric acid.—Nitrate of potash, Saltpetre, KO, NO_5. Exists native, being obtained in various places by the lixiviation of the soil. Formed artificially from the decomposition of organic matter in contact with bases; a colorless, transparent, crystalline salt; fuses at 600°, and, at a higher heat, evolves 39 per cent. of oxygen; a powerful oxidizing agent; sp. gr. about 2.00. Used as a source of nitric acid; as a manure; and in the manufacture of gunpowder.

a. Gunpowder. An intimate mixture of nitre, charcoal, and sulphur. Its projectile force, when fired, depends on the sudden evolution of a very large volume of gaseous matter from a small bulk of solid matter. Its goodness depends chiefly on the purity of the materials, and the accuracy of their mixture. In the explosion of powder, the nitre is decomposed by the charcoal, carbonic oxide and acid being produced, together with sulphurous acid and nitrogen; the sulphur is chiefly useful in increasing the rapidity of the action. In firing, one cubic inch of good powder produces about 790 cubic inches of gas; and as, at the moment of the explosion, this gas is red-hot, it is fair to reckon the expansion as about 1 into 2000. Gunpowder may be readily analyzed by boiling in water, which removes the nitre, and gentle heat, which sublimes off the sulphur. The ordinary composition is:—

	Nitre.	Charcoal.	Sulphur.
English—Waltham Abbey	75	15	10
French	75	12.5	12.5
United States	75	12.5	12.5
Prussia	75	12.5	12.5
China	75	14.4	9.9
Theoretical best	75	13.23	11.77

xvi. Potash and hyposulphurous acid.—Hyposulphite of potassa, KO, S_2O_2. Formed by the action of sulphurous acid on hydrosulphuret of potassa; a transparent, colorless, crystalline, deliquescent salt.

xvii. Potash and sulphyposulphuric acid. Sulphyposulphate of potassa, KO, S_3O_5. Obtained by the action of sulphur on a solution of bisulphite of potassa; a colorless, crystalline salt; not deliquescent.

xviii. Potash and sulphurous acid.—Sulphite of potassa, KO, SO_2. Formed direct; a white, crystalline salt. There is also a bisulphate, containing $KO, 2SO_2$.

xix. Potash and hyposulphuric acid.—Hyposulphate of potassa, $K O, S_2 O_5$. Formed when solutions of sulphate of potash and hyposulphate of baryta are mixed; a crystalline, soluble salt.

xx. Potash and sulphuric acid.—Sulphate of potassa, $K O, S O_3$. Formed direct; a white, crystalline, rather difficultly soluble salt; fusible at a red heat. Heated with excess of sulphuric acid, a bisulphate is formed, $K O, 2 S O_3$, which may be crystallized as $K O, 2 S O_3 H O$; this salt is the residue of the distillation of nitre with sulphuric acid, being then mixed with a sesquisulphate, $2 K O, 3 S O_3, H O$.

xxi. Potash and hypophosphorous acid.—Hypophosphite of potassa, $K O, P O, H O$. Formed direct; a very deliquescent, soluble, colorless salt.

xxii. Potash and phosphorous acid.—Phosphite of potassa, $K O, P O_3, H O$. Formed direct; a white, deliquescent, soluble salt.

xxiii. Potash and phosphoric acid.—Phosphates of potassa.

a. The common phosphate is obtained by the action of fused phosphoric acid on carbonate of potassa; it consists of $3 K O, P O_5$. There is also another tribasic salt, sometimes called a biphosphate, which consists of $K O, 2 H O, P O_5$.

b. The pyrophosphate. Obtained by the action of heat on the common phosphate; it does not crystallize, and contains $2 K O, P O_5$.

c. The metaphosphate. Formed by heating the biphosphate; it consists of $K O, P O_5$.

xxiv. Potash and selenious acid.—Selenite of potassa, $K O, Se O_2$. Formed direct; a white, crystalline, deliquescent salt. There is also a crystalline biselenite, $K O, 2 Se O_2$, and an uncrystallizable quadriselenite, $K O, 4 Se O_2$.

xxv. Potash and selenic acid.—Seleniate of potassa, $K O, Se O_3$. Formed direct, or by the action of selenium on fused nitre; a white, crystalline salt.

xxvi. Potash and boracic acid.—Borate of potassa, $K O, B O_3$. Formed by fusing anhydrous boracic acid with dry carbonate of potassa; a white, crystalline salt. There is also a biborate, $K O, 2 B O_3$ a very fusible, soluble, crystalline salt; formed direct;—and a sexborate, $K O, 6 B O_3$; formed when the preceding salts are digested with excess of boracic acid.

xxvii. Potash and silicic acid.—Silicate of potassa, $K O, Si O_3$. Formed direct by fusion, or digestion with water; a soluble, crystalline salt. There is also a bisilicate, a quatersilicate, and an octosilicate. Silicates of potash exist native in many rocks, associated with other silicates. They are also artificially prepared in the manufacture of glass—manufacture of artificial sandstone.

xxviii. Potash and fluosilic acid form silicofluoride of potassium, &c.

xxix. Potash and croconic acid.—Croconate of potassa, $K O, C_5 O_4$. Formed during the manufacture of potassium, and by the action of carbonic oxide on potassium; a yellow, crystalline salt.

xxx. Potash and rhodizonic acid.—Rhodizonate of potassa, $3 K O, C_7 O_7$. Formed by the action of water on a gray substance, composed of $K_3 C_7 O_7$; formed during the potassium process; a red salt; insoluble in water, or nearly so.

xxxi. Potash and oxalic acid.—Oxalate of potassa, $K O, C_2 O_3 + H O$. Formed direct; a white, crystalline salt, soluble in water.—There is also a binoxalate, which may be formed direct; it exists in the juice of the *Oxalis, Rumex,* and many other plants; hence called the salt of sorrel: it consists of $K O, H O, 2 C_2 O_3 + 2 H O$. When this salt is digested in acids, half its base is removed, and a quadroxalate is left, $K O, 4 C_2 O_3 + 7 H O$.

xxxii. Potash and carbonic acid.—Carbonate of potash, Pearlash, $K\,O, C\,O_2$. One of the most important of the salts of potash; because, as the acid can be expelled by any other acid, it may be conveniently employed as a source of all the salts of the alkali. Formed direct, or when potash is exposed to the air; obtained from wood ashes, by lixiviation and evaporation; a white, deliquescent, fusible, highly alkaline solid; very soluble in water, and crystallizing from its solution with $2\,H\,O$; insoluble in alcohol; sp. gr. 2.24. When supersaturated with carbonic acid, a bicarbonate is formed, $K\,O, 2\,C\,O_2 + H\,O$; a soluble, crystalline salt, which when heated reverts to the state of carbonate. There is also an intermediate carbonate, called the sesquicarbonate, composed of $2\,K\,O, 3\,C\,O_2$, which crystallizes with $12\,H\,O$, but this is perhaps only a mixture of the carbonate and bicarbonate.

xxxiii. Potash and sulphonaphthalic acid.—Sulphonaphthalate of potassa, $K\,O, C_{20}\,H_8, 2\,S\,O_3$. Formed direct; a white, pearly, crystalline salt, soluble in water and alcohol; heated it burns, and leaves sulphate of potassa.

xxxiv. Potash and cyanic acid.—Cyanate of potassa, $K\,O, N\,C_2\,O$. Formed direct, or by the action of heat on ferrocyanide of potassium; a white, fusible salt; decomposed by water, soluble in alcohol.

xxxv. Potash and fulminic acid. There is no fulminate of potassa; but the acid unites with potash and another base to form double salts.

xxxvi. Potash and cyanuric acid. Cyanurate of potassa, $K\,O, 2\,H\,O, 3\,N\,C_2$, or $2\,K\,O, H\,O, 3\,N\,C_2$; there is no salt with $3\,K\,O$.

xxxvii. Potash and hydrocyanic acid form cyanide of potassium, &c.

xxxviii. Potash and hydrosulphocyanic acid form sulphocyanide of potassium, &c.

xxxix. Potash and pectic acid.—Pectate of potassa, $K\,O$,

$C_{12} H_8 O_{10}$. Formed direct; a white, gum-like, soluble, uncrystallizable compound.

xl. Potash and mucic acid. — Mucate of potassa, $2 K O, C_{12} H_8 O_{14}$. Formed direct; a white, soluble salt.

xli. Potash and pyromucic acid.—Pyromucate of potassa, $K O, C_{10} H_3 O_5$. Formed direct; a white, soluble, deliquescent salt.

xlii. Potash and sulphovinic acid. — Sulphovinate of potassa, $K O, C_4 H_4, 2 S O_3$. Formed direct, and crystallizes with $H O$; a white, pearly, crystalline salt; very soluble in water.

xliii. Potash and oxalovinic acid.—Oxalovinate of potassa, $K O, C_4 H_5 O_2, C_2 O_3$. Formed when alcoholic solutions of potassa and oxalic ether are mixed; a white, crystalline salt; soluble in water, but almost insoluble in alcohol.

xliv. Potash and aldehydic acid.—Aldehydate of potassa, $K O, C_4 H_3 O_2$. Formed by the action of potassium on acetic ether; a white compound.

xlv. Potash and acetic acid.—Acetate of potassa, $K O, C_4 H_3 O_3$. Formed direct; exists in the juice of many plants; a white, crystalline, deliquescent salt; fusible, and readily soluble in water. With excess of acetic acid, it forms a white, crystalline, deliquescent binacetate, $K O, H O, 2 C_4 H_3 O_3$.

xlvi. Potash and chloracetic acid.—Chloracetate of potassa, $K O, C_4 Cl_3 O_3$. Formed direct; a white, silky, crystalline, deliquescent salt; decomposed by heat, with a feeble explosion.

xlvii. Potash and metacetonic acid.—Metacetonate of potassa, $K O, C_6 H_5 O_3$. Formed by the action of hydrate of potash on sugar.

xlviii. Potash and sulphomethylic acid.—Sulphomethylate of potassa, $K O, C_2 H_3 O_2, S O_3$. Formed direct, or by the action of sulphate of potash on sulphomethylate of baryta; a white, crystalline salt.

xlix. Potash and formic acid. Formiate of potassa,

KO, C_2OH_3. Formed direct; a white, difficultly crystallizable, deliquescent salt.

l. Potash and valerianic acid.—Valerianate of potassa, $KO, C_{10}H_9O_3$. Formed direct; a white, uncrystallizable, deliquescent, and very soluble salt.

li. Potash and kakodylic acid.—Kakodylate of potassa, $KO, C_4H_6A_sO_3$. Formed direct; a white, gum-like, uncrystallizable salt.

lii. Potash and benzoic acid.—Benzoate of potassa, $KO, C_{14}H_5O_3$. Formed direct; an efflorescent, white, crystalline salt. There is also a bibenzoate.

liii. Potash and salicylous acid.—Salicylite of potassa, $KO, C_{14}H_6O_4$. Formed direct; a white, pearly, crystalline salt, either anhydrous, or with $2HO$. There is also a bisalicylite of potassa, which crystallizes with HO.

liv. Potash and salicylic acid.—Salicylate of potassa, $KO, C_{14}H_5O_5$. Formed direct; a white, soluble, crystalline salt.

lv. Potash and aconitic acid.—Aconitate of potassa, KO, C_4HO_3. Formed direct, and not crystallizable.

lvi. Potash and amygdalic acid.—Amygdalate of potassa, $KO, C_{40}H_{26}O_{24}$. Formed direct, a soluble salt.

lvii. Potash and aspartic acid. Aspartate of potassa, $KO, C_8H_5O_6N$. Formed by boiling asparagine with solution of potassa; a very deliquescent, uncrystallizable, colorless salt.

lviii. Potash and butyric acid.—Butyrate of potassa, $KO, C_8H_8O_4$. Formed direct; a difficultly crystallizable and deliquescent salt.

lix. Potash and carbozotic acid.—Carbozotate of potassa, $KO, C_{12}H_2\ 3NO_4$. Formed direct; a brilliant, yellow, crystalline salt. When heated it fuses, and explodes with some violence.

lx. Potash and citric acid.—Citrate of potassa, $3KO$,

$C_{12}H_5O_{11}$. Formed direct; a deliquescent, crystalline salt. There is also another crystalline citrate, $KO, 2HO, C_{12}H_5O_{11}$; and a third salt, which is uncrystallizable, $2KO, HO, C_{12}H_5O_{11}$.

lxi. Potash and itaconic acid.—Itaconate of potassa, $KO, C_5H_2O_3$. Formed direct; a soluble, deliquescent, and uncrystallizable salt.

lxii. Potash and gallic acid.—Gallate of potassa, $KO, 3C_7H_3O_5$. Formed direct; a white, crystalline salt.

lxiii. Potash and ellagic acid. Ellagate of potassa, $KO, C_7H_2O_4$. Formed direct; a difficultly soluble, crystalline salt.

lxiv. Potash and hippuric acid.—Hippurate of potassa, $KO, C_{18}H_8O_5N$. Formed direct, and may be obtained crystallized with $2HO$. There is also a bisalt of $KO, HO, C_{18}H_8O_5N + 2HO$.

lxv. Potash and malic acid.—Malate of potassa, $2KO, C_8H_4O_8$. Formed direct; an uncrystallizable salt. There is also a bisalt, $KO, HO, C_8H_4O_8$.

lxvi. Potash and margaric acid.—Margarate of potassa, $KO, C_{34}H_{33}O_3$. Formed by boiling margaric acid with a solution of potassa, when the salt is deposited in brilliant, crystalline scales. There is also a bimargarate.

lxvii. Potash and oleic acid.—Oleate of potassa, $KO, C_{44}H_{40}O_4$. Formed like the margarate; a gelatinous, deliquescent soap. There is also a gelatinous binoleate, which is very difficultly soluble in water.

lxviii. Potash and silvic acid.—Silvate of potassa, $KO, C_{40}H_{30}O_4$. Formed direct; an uncrystallizable, soluble, resinous solid; the bisilvate is crystallizable.

lxix. Potash and racemic acid.—Racemate of potassa, $KO, C_4H_2O_5$. Formed direct; a transparent, colorless, efflorescent, crystalline salt. The biracemate, $KO, HO, 2C_4H_2O_5$, is also crystalline.

lxx. Potash and suberic acid. — Suberate of potassa, $KO, C_8 H_6 O_3$. Formed direct; a colorless, crystallizable, soluble, slightly deliquescent salt.

lxxi. Potash and succinic acid.—Succinate of potassa, $KO, C_4 H_2 O_3$. Formed direct; a crystalline salt. There is also a bisalt containing $KO, HO, 2 C_4 H_2 O_3$, which crystallizes with $4 HO$, and a quadrisuccinate, $KO, 4 C_4 H_2 O_3 + 5 HO$, or $KO, 3 HO, 4 C_4 H_2 O_3 + 2 HO$.

lxxii. Potash and tannic acid.—Tannate of potassa, $KO, C_{18} H_8 O_{12}$. Formed direct; an earthy-looking, white solid.

lxxiii. Potash and tartaric acid.—Tartrate of potassa, $2 KO, C_8 H_4 O_{10}$. Formed direct, or by boiling the bitartrate with potassa; a colorless, transparent, soluble, slightly deliquescent salt. The bitartrate, which is formed in wine, being deposited as Argol or tartar, is a white, crystalline salt, almost insoluble in water; it consists of $KO, HO, C_8 H_4 O_{10}$; its sp. gr. is 1.95; when burnt it leaves "back flux," a mixture of carbonate of potassa and charcoal, used in metallurgy.

lxxiv. Potash and uric acid.—Urate of potassa, $2 KO, C_{10} H_2 O_4 N_4$. Formed direct; a crystalline salt, which, by the action of carbonic acid, passes into a binurate, $KO, HO, C_{10} H_2 O_4 N_4$; a white, granular salt.

D. Potassium and chlorine.—Chloride of potassium, KCl. The metal burns brightly in chlorine; formed also by the action of hydrochloric acid on potassa or its carbonate; a white, crystalline salt; fusible at a bright red heat, and volatile at a white heat; its vapor is decomposed by steam.

E. Potassium and iodine.—Iodide of potassium, KI. Formed direct, heat and light being evolved; or by heating the iodate, or by the action of carbonate of potassa on iodide of zinc, KO, CO_2 and ZnI become KI and ZnO, CO_2; a white, crystalline solid; easily soluble in water; fusible at a red heat, and slightly deliquescent in damp air. The aqueous solution dissolves a considerable quantity of iodine.

F. Potassium and bromine.—Bromide of potassium, K Br. Formed like the iodide; a colorless, crystalline solid; soluble in water, and fusible at a red heat; sp. gr. 2.4.

G. Potassium and fluorine.—Fluoride of potassium, K Fl. Obtained by evaporating a solution of potassa in hydrofluoric acid to dryness; a white, difficultly crystallizable, deliquescent solid.

H. Potassium and hydrogen.—Hydruret of potassium. When the metal is heated in the gas it absorbs a portion, and forms an infusible gray compound. There is also a spontaneously inflammable, gaseous compound, potassiuretted hydrogen, formed when potassium acts upon water.

I. Potassium and amidogen.—Potassiamide, K—NH_2. Formed by heating potassium in ammonia; hydrogen is evolved, and an olive-colored solid formed; it fuses at 212°, deliquesces in the air, and is resolved into potassa and ammonia by water. When heated out of contact of air a gray solid is formed, which appears to consist of K N, ammonia being evolved.

J. Potassium and sulphur.—Sulphurets of potassium.

i. Protosulphuret of potassium, K S. Formed direct, out of contact of air; or by igniting sulphate of potassa with charcoal powder; a brownish-gray solid; fusible below redness; deliquescent, and soluble in water, when it forms a solution of hydrosulphuret of potassa, K O, S H. This solution may also be formed from sulphuretted hydrogen and solution of potassa, and then yields crystals of the hydrosulphuret.

ii. Bisulphuret of potassium, K S_2. Formed direct, by heating proper proportions of sulphur and potassium. It forms a yellow solution of hydrosulphuret of potassa in water, K O, S_2 H.

iii. Tersulphuret of potassium, K S_3. Formed by heating dry carbonate of potassa with half its weight of sulphur. Compounds of $K_2 S_7$—$K S_4$ and $K_2 S_9$ have also been obtained.

iv. Pentasulphuret of potassium, $K S_5$. Obtained by passing sulphuretted hydrogen over ignited sulphate of potassa.—The "Liver of Sulphur," obtained by fusing sulphur with potash or its carbonate, is a mixture of some of the preceding compounds.

K. Potassium and phosphorus.—Phosphuret of potassium, K P. Obtained direct, out of contact of air; decomposed by water, phosphuretted hydrogen being evolved.

L. Potassium and selenium.—Seleniuret of potassium, K Se. They combine direct, with strong action; a gray, crystalline solid; soluble in water, and resembling the sulphuret.

M. Potassium and fluoboric gas.—Fluoboride of potassium, $K B F_4$. Obtained by adding fluoboric gas to a solution of fluoride of potassium; a white, pulverulent, crystalline solid; it probably consists of K F, $B F_3$.

N. Potassium and fluosilicic gas.—Silicofluoride of potassium, K F, Si F_2. Obtained by passing fluosilicic acid gas into a solution of potassa; a white, crystalline, pulverulent solid; gelatinous when first formed in the solution of potassa; it is sparingly soluble in boiling water.

O. Potassium and carbon.—Carburet of potassium, K C. Formed during the process for preparing potassium; a black solid, spontaneously inflammable.

P. Potassium and cyanogen. Cyanide of potassium, K C_2 N. Formed direct, or by evaporating a solution of potassa in hydrocyanic acid; sometimes formed in blast furnaces; formed by passing nitrogen over ignited carbonate of potassa and charcoal; a white, crystalline salt; fusible below a red heat; soluble in water; heated in the air it becomes cyanate of potassa; acted on by acids, it yields hydrocyanic acid.

Q. Potassium and mellone.—Mellonide of potassium, K $C_6 N_4$. Obtained direct, or by dissolving mellone in solution of potassa; a yellowish, opaque, semi-crystalline solid, which crystallizes with 5 H O (?), in colorless, transparent needles.

R. Potassium and sulphocyanogen. — Sulphocyanide of potassium, K C_2 N S_2. Obtained by boiling sulphur in a solution of the cyanide of potassium, or by heating the ferrocyanide of potassium with sulphur and carbonate of potassa; a white, deliquescent, crystalline solid; fusible in close vessels below redness; decomposed when heated in the air; a useful test for iron.

S. Potassium and seleniocyanogen.—Seleniocyanide of potassium, K N C_2 Se_2. Obtained by heating ferrocyanide of potassium with selenium; a colorless, transparent, crystalline solid.

T. Tests.—Salts mostly deliquescent; color flame purplish white; forms a double chloride with platinum; difficultly soluble bitartrate; forms alum with sulphate of alumina.

U. Uses.—Potassium, used in the laboratory as a deoxidizer; potash ditto, as a powerful base; chlorate used for matches and fireworks; nitre used for gunpowder; carbonate used in soap and glass-making; chloride and sulphate used in alum-making.

2. Sodium. Natrium. Symb. Na. Eq. 24. (23.62.)

A. Exists as an oxide in many rocks; occurs in marine plants, and is obtained from their ashes; occurs as a chloride, common salt. Obtained from the oxide like potassium.

B. Properties.—A brilliant, silvery-white metal; very oxidizable; must be preserved in naphtha; soft like wax at common temperature, and does not become brittle even at 32°; fuses at 190°; and boils at a red heat, forming a colorless vapor. Sp. gr. 0.9348; when heated it burns in the air or oxygen with a bright yellow flame. It does not burn on cold water, but it burns readily on hot water, on thick viscid liquids, and on ice.

C. Sodium and oxygen.—Oxide of sodium, pure soda, $Na\,O = 24 + 8 = 32$. Formed direct; obtained by strongly heating the hydrated oxide with sodium; resembles potash; combines with water to form a hydrate, caustic soda, $Na\,O, H\,O$; obtained by the action of lime on carbonate of soda. A white, semi-transparent, brittle, deliquescent solid; sp. gr. 2.0; fusible at a red heat, and volatile at a white heat; the solution is strongly alkaline. Soda unites with acids to form salts. There is also a suboxide, $Na_2\,O$, and a peroxide, $Na\,O_2$, like those of potassium.

i. Soda and hypochlorous acid.—Hypochlorite of soda, $Na\,O, Cl\,O$. Formed by the action of chlorine on solution of soda, or its carbonate; evolves chlorine when exposed to the air, and hence used as a disinfecting and bleaching agent.

ii. Soda and chloric acid.—Chlorate of soda, $Na\,O, Cl\,O_5$. Formed like the chlorate of potassa, which it closely resembles.

iii. Soda and iodic acid.—Iodate of soda, $Na\,O, I\,O_5$. Formed like the potash salt; crystallizes with $2\,H\,O$.

iv. Soda and nitric acid.—Nitrate of soda, $Na\,O, N\,O_5$. Cubic nitre. Exists native abundantly in Peru; may be formed direct. A colorless, slightly deliquescent salt; fusible at about 600°; it is used as a source of nitric acid, and as a manure; but, in consequence of its hygrometric nature, it cannot be used for gunpowder, except in very dry climates. Its solution is decomposed by carbonate of potassa, nitrate of potassa and carbonate of soda being formed.

v. Soda and hyposulphurous acid.—Hyposulphite of soda, $Na\,O, S_2\,O_2$. Obtained by adding carbonate of soda to hyposulphite of lime; it crystallizes with $5\,H\,O$, and is slightly deliquescent.

vi. Soda and sulphuric acid.—Sulphate of soda, Cheltenham or Glauber salts, $Na\,O, S\,O_3$. Formed direct, or by the action of sulphuric acid on common salt, or the nitrate of soda; it exists native in Spain, Switzerland, and in various mineral springs, &c. A white, crystalline solid, fusible at a dull red heat; easily soluble in water, and forming crystals containing 8 H O and 10 H O. They are efflorescent. There is also a bisulphate, $NaO, 2SO_3$, which may be formed direct, and a sesquisulphate 2 Na O, $3SO_3$. Sulphate of soda combines with sulphate of ammonia, to form a crystalline, double salt; it is best obtained by adding ammonia to the bisulphate, and consists of $NaO, NH_3, HO, 2SO_3 + 4HO$.

vii. Soda and phosphoric acid.—Phosphates of soda.

a. Tribasic phosphates. By saturating common phosphoric acid, as it is obtained from bone earth, with carbonate of soda, the common rhombic phosphate is obtained, $2NaO, HO, PO_5 + 24HO$. A white, crystalline salt; sp. gr. 1.33; slightly efflorescent; fusible in its own water of crystallization; when acted on by excess of caustic soda, it forms a subphosphate, $3NaO, PO_5 + 24HO$; when acted on by excess of phosphoric acid, it forms biphosphate, $NaO, 2HO, PO_5 + 2HO$. Phosphate of soda combines with ammonia, to form a curious double salt, the ammonio-phosphate, or microcosmic salt; it exists in urine, and may be formed direct; it consists of $NaO, NH_3, 2HO, PO_5 + 8HO$.

b. Bibasic or pyrophosphates. The pyrophosphate is obtained by heating the common phosphate to redness; it consists of $2NaO, PO_5$, and crystallizes with 10 H O; by carefully heating the biphosphate of soda, a bipyrophosphate is obtained, which consists of NaO, HO, PO_5.

c. Monobasic phosphate; obtained by igniting the biphosphate of soda; it consists of NaO, PO_5.

viii. Soda and boracic acid.—Biborate of soda, Tinkal, $NaO, 2BO_3$. Exists native in India, Persia, and Thibet; formed artificially direct; crystallizes with $10HO$, forming slightly efflorescent, colorless, transparent crystals. Heated, it parts with its water of crystallization, and fuses at a red heat. It is used as a flux; in the composition of pastes or artificial gems; and by workers in metal, soldering, &c. Borax may also be crystallized with $5HO$.

ix. Soda and silicic acid.—Silicate of soda, NaO, SiO_3. Formed direct by fusion. There are several silicates which occur native in minerals, and are formed artificially by glass-makers, entering into the composition of the best kinds of glass. The tersilicate, $NaO, 3SiO_2$, is a soluble crystalline salt. There is also a hydrated quatersilicate, composed of $NaO, 4SiO_2 + 12HO$.

x. Soda and carbonic acid.—Carbonate of soda, the "soda" of commerce, NaO, CO_2. Formed direct; obtained from the ashes of marine plants, constituting kelp, barilla, and varec; manufactured from the sulphate by igniting it with charcoal, which deoxidizes it, and forms a sulphuret; this is decomposed by carbonate of lime, which yields an impure carbonate of soda; from this the pure carbonate is manufactured; obtained also by the action of hot steam on mixtures of sulphate of soda, or salt, with alumina at a bright red heat, &c. Carbonate of soda crystallizes with HO—$5HO$—$6HO$—$8HO$, and with $10HO$,—the latter being the composition of the salt met with in commerce; exposed to air it effloresces. It is much used in the arts, particularly in glass and soap making. It is useful as a source of all the other soda salts. There is also a native carbonate of soda, Trona, which is found in Africa, India, South America, &c. It consists of $2NaO, 3CO_2$, and crystallizes with

3 H O and 4 H O. This salt may be obtained artificially by boiling a solution of the bicarbonate. The bicarbonate of soda may be formed by exposing the carbonate to the action of excess of carbonic acid. It consists of Na O, H O, 2 C O_2. This salt is chiefly employed in medicine. The analysis of the carbonates of soda, and also those of potash, forms a special branch of analysis, termed alkalimetry; the most convenient process consists in measuring the quantity of sulphuric acid, of a known strength, which a given quantity of the alkaline salt can saturate.

xi. Soda and acetic acid.—Acetate of soda, Na O, C_4, H_3 O_3. Formed direct; a colorless salt; soluble in water, and crystallizing with 6 H O. It is prepared from crude, pyroligneous acid, as a source of the pure acid; and, ignited, leaves a mixture of soda and charcoal, very convenient as a source of sodium.

D. Sodium and chlorine.—Chloride of sodium, common salt, Na Cl. It occurs abundantly native as rock salt, and dissolved in sea water, whence it is obtained pure by evaporation and crystallization. Sodium burns in chlorine. A white, crystalline solid; not deliquescent when pure; decrepitates when heated, from unequal expansion; fusible at a red heat; volatile at a white heat; decomposed at a white heat by steam; sp. gr. 2.125; readily soluble in water; insoluble in alcohol; by the action of sulphuric acid, it yields muriatic acid and sulphate of soda; heated with sulphuric acid and oxide of manganese, it yields chlorine. Extensively used as a condiment, and essential in the processes of cookery, and the preservation and preparation of food; valuable as a source of soda and its salts; used in the manufacture of soda, muriatic acid, chlorine, &c. Employed in agriculture as a manure; sometimes used as a flux, and as a glaze in the manufacture of earthenware.

E. Sodium and iodine.—Iodide of sodium, Na I. Exists

in sea water and the ashes of marine plants; formed easily direct; and resembles the iodide of potassium, the chief source of iodine.

F. Sodium and bromine.—Bromide of sodium, Na Br. Exists in sea water; formed easily direct; decomposed by chlorine; the chief source of bromine.

G. Tests.—Salts mostly efflorescent; color, flame yellow; forms a soluble double chloride with platinum, and a soluble bitartrate.

H. Uses.—The metal as a deoxidizer; soda as a base; the nitrate as sources of nitric acid, and as manure; the sulphate in medicine; the carbonate as a detergent, and manufacture of soap and glass; the chloride as a source of the other soda salts; as manure; in metallurgy; as a source of chlorine compounds, &c.

3. Lithium. Symb. L. Eq. 7. (6.43.)

A. Exists as an oxide in certain very rare stones, Spodumene, Petalite, Lepidolite, &c.; obtained from its hydrated oxide by the action of electricity. A silvery, white, very oxidizable metal, burning on exposure to the air, and forming lithia.

B. Lithium and oxygen.—Lithia, oxide of lithium, Li O. Formed direct; obtained by igniting the acetate of lithia. A white, caustic, alkaline solid, less soluble in water than potassa or soda.

i. Salts of lithia. It combines with acids to form a class of salts in many respects resembling those of soda. They are mostly soluble in water, crystalline and deliquescent. The phosphate is insoluble, and so is the carbonate.

C. Lithium and chlorine.—Chloride of lithium, Li Cl. A deliquescent, crystallizable compound; formed by dissolving lithia or its carbonate in muriatic acid, and evaporating to dryness out of contact of air.

D. Tests.—Easily distinguished from the two other fixed alkalies by its superior saturating power for acids, 15 parts being equal to 32 of soda and 48 of potassa;

the insolubility of its carbonate, and the deep red color which it gives to flame.

4. CALCIUM. Symb. Ca. Eq. 20.

A. Exists abundantly as an oxide combined with different acids; the metal is obtained from its hydrated oxide by the action of electricity. A silvery, white, very oxidizable metal.

B. Calcium and oxygen.—Oxide of calcium, Lime, Ca O. Obtained by heating its carbonate, as in the process of lime burning. A grayish, white solid; sp. gr. 3.08; very difficultly fusible; slowly volatile at a white heat; remarkably luminous when ignited; hence used in the Drummond light. It is a distinctly alkaline earth; is caustic and acrid; reddens vegetable yellow tests; is a powerful base; exposed to air it absorbs water rapidly, and forms a hydrate; moistened with water it swells, becomes very hot, and falls into a dry powder, which is the hydrate or slaked lime; one part of lime requires 750 parts of water to dissolve it at 60°, and 1280 at 212°. The hydrate of lime crystallizes as Ca O, H O. Lime water is useful as a test for carbonic acid. There is also a crystalline peroxide of calcium, $Ca\,O_2$. Lime is used in consequence of its powerful basic properties in the preparation of potassa, soda, &c.; it is employed in agriculture, the manufacture of mortar and cements, in metallurgy as a flux, &c.

C. Salts of lime.

i. Lime and hypochlorous acid.—Hypochlorite of lime, bleaching powder, Ca O, Cl O (?). Obtained by the action of chlorine on the hydrate of lime. A white powder, having a faint odor of chlorine, partially soluble in water, and readily decomposed by acids which liberate chlorine; it is manufactured on a large scale as a bleaching agent for calico printers, paper makers, &c.; it is also used as a disinfecting material, and contains on an average about 30 per cent. of chlorine.

ii. Lime and nitric acid.—Nitrate of lime, Ca O, NO_5. Formed direct; occurs native in various calcareous soils, and in old mortar. A white, soluble, deliquescent salt; it may be decomposed by carbonate of potassa, and is sometimes employed as a source of nitre.

iii. Lime and sulphuric acid.—Sulphate of lime, gypsum, selenite, Ca O, $S O_3$. Easily formed direct; exists abundant native as Plaster-stone, Alabaster, &c. A white, opaque, earthy solid; transparent, and crystallized with 2 H O; one part dissolves in 500 parts of water; the sp. gr is 2.927 to 2.970. When the native hydrated sulphate is heated to 400°, it loses its 2 H O, and becomes Plaster of Paris, the setting of which depends on its re-absorbing this water. Sulphate of lime frequently exists in water, and renders it hard and unfit for household uses. Sulphate of lime is a valuable manure.

iv. Lime and phosphoric acid.—Phosphates of lime. The tribasic phosphate, 3 Ca O, $P O_5$, occurs native in Spain and other places, as Apatite, &c.; it exists abundantly in bone, and may be obtained from bone-ash, or formed direct; it is a white solid, almost insoluble in water, though more soluble in water containing carbonic acid; the native varieties are phosphorescent when heated; used as a source of phosphoric acid, and as a manure. The other phosphates may readily be formed from this salt: by acting on bone earth by sulphuric acid, a superphosphate of lime is obtained, which is much used as a manure; as sold, it appears to consist of phosphoric acid, phosphate of lime, and sulphate of lime.

v. Lime and silicic acid.—Silicates of lime. Silicates of lime exist native, as Table-spar and other minerals; associated with iron, &c., they exist in Hornblende, Pyroxene, Stilbite; they may also be formed direct; and are produced in many metallurgical processes in which lime is used as a flux.

vi. Lime and boracic acid.—Borate of Lime, Ca O, 2 B O_3. May be formed direct; and exists native as Botryolite; it is almost insoluble in water, and fuses at a bright red heat.

vii. Lime and oxalic acid.—Oxalate of lime, Ca O C_2 O_3. Formed direct, or by adding oxalate of ammonia to any solution containing lime, even as a sulphate. An insoluble, white salt, decomposed by ignition, which leaves pure lime.

viii. Lime and carbonic acid.—Carbonate of lime, Ca O, C O_2. Occurs native abundantly, as Chalk, Marble, Limestone, Kunkur, Calcareous spar, &c.; may be formed by exposing hydrate of lime to carbonic acid; white; insoluble in water; tasteless; has no action on tests; is infusible; loses C O_2 when heated; it is easily decomposed, like all other carbonates, by any acid except the hydrocyanic; dissolves in water containing excess of carbonic acid; exists in many natural waters thus dissolved, rendering them hard, and precipitating from them in the solid state as a "fur," when boiled. Used as a manure; as a source of quicklime; as a flux; for mortars, which consist chiefly of carbonate of lime; the hydrate of lime which they contain gradually hardening, as it re-absorbs carbonic acid from the air, and passes again into the state of carbonate.

ix. Lime and citric acid.—Citrate of lime, Ca O, C_{12} H_5 O_{11}. Formed direct; a white, soluble salt. It is manufactured from crude lime or lemon juice, and used as a source of pure citric acid. It may be obtained, crystallized with H O, and also with 4 H O.

D. Calcium and chlorine, Ca Cl. Formed by heating lime in chlorine; or dissolving lime or its carbonate in muriatic acid, and evaporating to dryness; it exists native in sea water, and some mineral springs. A grayish, white, translucent solid, fusible at a red heat, very deliquescent, and hence used to dry gases, &c. It is very soluble in water, and forms crystals with 6 H O, commonly called muriate of lime.

E. Calcium and fluorine.—Fluoride of calcium, Fluor spar, Ca Fl. May be formed artificially, by acting on lime by hydrofluoric acid; exists native abundantly, crystallized as fluor spar. A hard solid; sp. gr. 3.0; infusible, but useful as a flux; insoluble in water; some of the varieties phosphoresce when heated. Fluoride of calcium is frequently found in fossil bones; it is interesting, as almost the only source of the fluorine compounds.

F. Calcium and phosphorus.—Phosphuret of calcium, Ca P. Formed by passing the vapor of phosphorus over red-hot lime. A brown, earthy-looking solid. It is decomposed by water, spontaneously inflammable, phosphuretted hydrogen being evolved.

G. Tests.—The soluble salts are all decomposed by the carbonated alkalies and by potassa or soda, but not by ammonia; they are also all precipitated by oxalate of ammonia, the affinity of lime for oxalic acid being very strong.

5. Barium. Symb. Ba. Eq. 69. (68.64.)

A. Exists in nature as an oxide, combined with sulphuric and carbonic acid; far less abundant than lime; metal obtained by passing the vapor of potassium over the ignited oxide of barium. An iron-gray solid; sp. gr. 2.0; very oxidizable; burns when heated in the air with a red light, forming an oxide.

B. Barium and oxygen.—Oxide of barium, baryta, Ba O. Obtained by igniting iodate or nitrate of baryta. A gray, earthy-looking substance; sp. gr. about 4.0; has a strongly alkaline reaction, an acrid taste, and, as well as nearly all of its compounds, is poisonous. It absorbs water eagerly, like quicklime, and forms a white hydrate, Ba O, H O; this readily dissolves in water, forming baryta water. Baryta is a powerful base, and has a strong affinity for acids. Barium also forms a peroxide, Ba O_2; which may be obtained by passing oxygen over ignited baryta; this compound is chiefly interesting as a source of the peroxide of hydrogen, when acted on by hydrochloric acid.

C. Baryta and acids.

i. Baryta and chloric acid.—Chlorate of baryta, Ba O, Cl O_5. Formed by acting on baryta water by chlorine. A colorless, soluble, crystalline salt, which yields chloric acid when decomposed by sulphuric acid.

ii. Baryta and nitric acid.—Nitrate of baryta, Ba O, N O_5. Formed by dissolving carbonate of baryta, or sulphuret of barium, in nitric acid; a white, crystalline, soluble salt; insoluble in nitric acid; used as a test for sulphuric acid.

iii. Baryta and sulphuric acid.—Sulphate of Baryta, Ba O, S O_3. Exists abundantly native as Cawk, or heavy spar; may be formed direct, or by adding any soluble sulphate to a solution containing baryta. A white salt; sp. gr. 4.70; insoluble in water and acids. Prepared artificially as a white pigment, or permanent white, not being liable to blacken like white lead; used also in porcelain making; is not poisonous; it is decomposed at a white heat by steam.

iv. Baryta and carbonic acid.—Carbonate of baryta, Ba O, C O_2. Exists native as Witherite; formed direct. A white, insoluble salt; sp. gr. 4.33; it has a slightly alkaline reaction; and is eminently poisonous. The salts of baryta are easily prepared from the carbonate; they are useful as means of obtaining many acids pure; as the baryta salts are all completely decomposed by the careful addition of sulphuric acid, which precipitates the whole of the earth as an insoluble sulphate, and leaves the acid, with which it was previously combined, pure.

D. Barium and chlorine, Ba Cl. Obtained by passing chlorine over ignited baryta; or by dissolving carbonate of baryta in hydrochloric acid, and evaporating to dryness; a grayish, white, translucent solid, fusible at a red heat; sp. gr. 3.80; it is easily soluble in water, and forms transparent, colorless crystals, containing 2 H O. Used chiefly as a test for sulphuric acid.

E. Barium and sulphur.—Sulphuret of barium, Ba S. Formed by passing sulphuretted hydrogen over ignited baryta; or by igniting a mixture of powdered sulphate of baryta and charcoal; it is soluble in water and easily decomposed by acids, or even mere exposure to the air.

F. Tests.—The soluble salts are all precipitated by sulphuric acid, and by a solution of sulphate of Strontia, and the precipitate is insoluble in nitric acid; they are also precipitated white by alkaline carbonates.

6. STRONTIUM. Symb. Sr. Eq. 44. (43.84.)

A. Exists native, like baryta, as a carbonate and sulphate, but less abundant. The metal is obtained like barium.

B. Strontium and oxygen.—Oxide of strontium, Strontia, Sr O. Obtained like baryta, which in most respects it resembles; moistened, it slakes, and forms a hydrate, which dissolves in water, forming an alkaline solution.

C. Strontia and acids.

i. Strontia and nitric acid.—Nitrate of strontia, Sr O, N O_5. Readily formed direct; a white salt, soluble in water, and crystallizing with 5 H O. Used in the manufacture of red fire and other pyrotechnic preparations; tinges the flame of alcohol red.

ii. Strontia and sulphuric acid.—Sulphate of strontia,

$Sr\,O, S\,O_3$. Occurs native as Cœlestine; may be formed direct; very sparingly soluble in water. It is converted into sulphuret when ignited with charcoal; and is decomposed at a red heat by steam.

iii. Strontia and carbonic acid.—Carbonate of strontia, $Sr\,O, C\,O_2$. Occurs native, though rare, as Strontianite; may be formed direct; a difficultly soluble, white compound.

D. Strontium and chlorine.—Chloride of strontium, Sr Cl. Formed by passing chlorine over ignited strontia; or by dissolving the carbonate in hydrochloric acid, and evaporating the solution to dryness. A gray solid, soluble in water and alcohol, and crystallizing as a colorless salt with 6 H O.

E. Tests.—Distinguished from baryta salts by the slight solubility of the sulphate, and by the red color which it gives to flame.

7. Magnesium. Symb. Mg. Eq. 12. (12.67.)

A. Exists native as an oxide, combined with carbonic acid, &c., and as a chloride in sea water. The metal is obtained by heating the chloride with potassium. A silvery, white, brilliant metal; hard, malleable, fusible at a red heat; slowly oxidized by water; and burning brightly in the air when heated.

B. Magnesium and oxygen.—Oxide of magnesium, magnesia, Mg O. Obtained by heating the carbonate. A white solid; sp. gr. 3.0; nearly insoluble in water; has a very feeble alkaline reaction; absorbs carbonic acid and moisture very slowly when exposed to the air; does not heat or slake when moistened with water, but forms a hydrate of Mg O + H O, which occurs native in New Jersey and in Scotland. The presence of magnesia in limestone very greatly modifies its nature; the quicklime obtained from it does not become mild when exposed to the air; hence it cannot be used so well as pure lime for agricultural purposes.

C. Magnesia and acids.

i. Magnesia and sulphuric acid.—Sulphate of magnesia, Epsom salts, $Mg\,O, S\,O_3$. Occurs native in South America, in various mineral springs, &c.; may be formed direct; manufactured from "bittern" of sea water, and from magnesian limestones. A white salt, very soluble in water, and crystallizing with 7 H O. It may also, by proper management, be crystallized with H O—2 H O—6 H O and 12 H O. Used chiefly in medicine. It forms double sulphates with ammonia, potassa, and soda, &c. The composition of the latter is $Na\,O, Mg\,O, 2\,S\,O_3 + 6\,H\,O$.

ii. Magnesia and phosphoric acid.—Phosphates of magnesia. By the addition of common phosphate of soda to a solution of sulphate of magnesia, a crystalline tribasic phosphate is obtained, $2\,Mg\,O, H\,O, P\,O_5 + 14\,H\,O$; when this salt is heated it yields the bibasic phosphate, $2\,Mg, P\,O_5$. The monobasic phosphate may be formed by adding monobasic phosphate of soda to a solution of acetate of magnesia, $Mg\,O, P\,O_5$. There is also a double phosphate of ammonia and magnesia which belongs to the tribasic class, and consists of $N\,H_3, 2\,Mg\,O, P\,O_5 + 13\,H\,O$. It is formed by adding phosphate of ammonia to a solution of any magnesian salt. This salt exists in urine, and is an excellent manure.

iii. Magnesia and silicic acid.—Silicate of magnesia, $2\,Mg\,O, Si\,O_3$. Occurs native as Olivin and Chrysolite, and other similar compounds, and also found native, associated with different silicates. They may be formed artificially direct.

iv. Magnesia and carbonic acid.—Carbonate of magnesia, $Mg\,O, C\,O_2$. Occurs native as Magnesite in North America, Piedmont, India, &c. Associated with carbonate of lime it forms Bitterspar, Dolomite, and magnesian limestone. Manufactured from solutions of the sulphate or chloride by precipitation with an alkaline carbonate; much used in medicine; a white, tasteless powder, insoluble in water, but soluble in water containing carbonic acid, and deposited from this solution crystallized with 3 H O. The pharmaceutical preparation is not pure proto-carbonate, but appears to consist of 5 Mg O, $4\,C\,O_2 + 6\,H\,O$.

D. Magnesium and chlorine.—Chloride of magnesium, Mg Cl. Obtained by passing dry chlorine over a mixture of magnesia and charcoal at a red heat; cannot be obtained by evaporating the muriate. A white, deliquescent, crystalline solid, soluble in water, and crystallizing with 6 H O. It exists naturally in sea water, and is obtained from it by evaporation, remaining in the Bittern or uncrystallizable residue after the separation of the common salt.

E. The salts of magnesia are mostly soluble; their solutions are not precipitated by ammonia where sal-ammoniac is present, but the magnesia is thrown down from such a solution by phosphoric acid.

Sect. II.—Metals of the Earths.

1. Aluminum. Symb. Al. Eq. 14. (13.69.)

A. Exists native abundantly as an oxide, in clay and many minerals; the metal is obtained by the action of potassium on chloride of aluminum. An iron-gray solid, having little metallic lustre; almost infusible; and combustible when heated in the air; it is not acted on by water.

B. Aluminum and oxygen.—Oxide of aluminum, Alumina, $Al_2\,O_3$. Exists native almost pure in the Sapphire, Ruby, Topaz, Corundum, &c.; obtained by adding ammonia to a solution of alum, or by evaporating an aqueous solution of chloride of aluminum. A white, tasteless solid, insoluble in water, and wholly without action on colored tests; fusible at a bright white heat; when moist and freshly precipitated it is remarkably plastic; it has a strong affinity for organic fibre, and also for coloring matter; hence of much importance in dyeing, and the manufacture of pigments; it combines with acids, but it is a feeble base. The hydrate occurs native as Diaspore, containing 2 H O; and as Gibbsite, containing 3 H O. The artificial gelatinous hydrate contains 8 H O, of which 5 equivalents are expelled by a heat of 212°, but the whole cannot be expelled altogether by heat; when, however, it has been precipitated from a concentrated solution it is not gelatinous but pulverulent, and then may be wholly dried by a red heat. Alumina dissolves in solutions of potassa and soda, forming definite compounds of K O, Al_2 O_3, and Na O, Al_2 O_3; they may be crystallized with 3 H O, and are easily decomposed by carbonic acid. Alumina and lime have also considerable affinity for each other; this compound may be formed direct; it is of importance as a component of several of the best mortars and cements; the value of Dorking lime depends chiefly on the presence of alumina.

C. Alumina and acids.

i. Alumina and sulphuric acid.—Sulphates of Alumina, Al_2 O_3, S O_3; occurs native with 9 H O as Websterite or Aluminite, in Sussex, &c.; and may be artificially prepared by adding ammonia to the tersulphate, Al_2 O_3, 3 S O_3, which is formed by digesting hydrate of alumina in sulphuric acid; a white, crystalline salt, containing 18 H O. When the sulphate is boiled in a solution of the tersulphate, a bisulphate is formed, Al_2 O_3, 2 S O_3; this is uncrystallizable. Sulphate of alumina forms a number of remarkable double salts; of these the most important is common Alum, or sulphate of potassa and alumina, K O, S O_3 + Al_2 O_3, 3 S O_3 + 24 H O. This is manufactured on a large scale from aluminous shales, which contain sulphuret of iron; they are burnt, and subsequently lixiviated; and the crude solution of sulphate of alumina thus obtained is mixed with sulphate of potash, and crystallized. A colorless, transparent, crystalline salt; soluble in water; fusible in its own water of crystallization when heated. Alum is largely used in the arts, chiefly in dyeing, and in the preparation of aluminous mordaunts. Other similar salts may be formed with ammonia, soda, &c.; they are analogous in composition, and crystallize with 24 H O.

ii. Alumina and phosphoric acid.—Phosphate of alumina, $3\ Al_2\ O_3,\ 2\ P\ O_5 + 12\ H\ O$; occurs native in clay slate as Wavellite; and the turquoise appears to consist of $2\ Al_2\ O_3,\ P\ O_5 + 5\ H\ O$, colored blue by copper. These phosphates may also be made artificially.

iii. Alumina and silicic acid.—Silicates of alumina. Many of these occur native as in Cyanite, $2\ Al_2\ O_3,\ Si\ O_3$, and associated with alkaline silicates in Felspar, Zeolite, &c. Artificial compounds are readily made; the basis of good pottery and china is a silicate of alumina; clays consist of silica and alumina, but are chiefly mixtures.

iv. Alumina and acetic acid.—Acetate of alumina, $Al_2\ O_3,\ 3\ C_4\ H_3\ O_3$. May be formed direct; manufactured by adding alum to acetate of lead; a gum-like, uncrystallizable, very soluble, deliquescent salt; it is much used in dyeing.

D. Aluminum and chlorine.—Chloride of aluminum, $Al_2\ Cl_3$. Formed by passing dry chlorine over an ignited mixture of alumina and charcoal; cannot be obtained by evaporating the muriatic solution of alumina. A yellowish-white, crystalline solid, which is highly deliquescent, and may be crystallized with 12 H O.

2. Glucinum. Symb. G. Eq. 7. (7.16.)

A. Exists as an oxide; very rare. Obtained by the action of potassium on chloride of glucinum.

B. Glucinum and oxygen.—Oxide of glucinum, glucina, G O. Exists in the Emerald Beryl, Euclase, &c.; in many respects resembles alumina; remarkable for the sweet taste of several of its salts.

3. Zirconium. Symb. Zr. Eq. 33. (33.60.)

A. Exists as an oxide; very rare. Obtained by the action of potassium on the potassi fluoride of zirconium; a black, or iron-gray solid, which burns when heated in the air.

B. Zirconium and oxygen.—Oxide of zirconium, Zirconia, Zr O. Exists in the zircon, and a few other very rare minerals; a white, infusible, insoluble substance; sp. gr. 4.3; brightly luminous, like lime, when ignited in the blowpipe flame; it combines with bases to form salts, most of which are uncrystallizable.

4. Yttrium. Symb. Y. Eq. 32. (32.20.)

A. Exists as an oxide in Gadolinite and Yttrocerite; very rare; obtained by acting on chloride of yttrium by potassium; a brittle, gray, almost infusible solid, which burns brilliantly in the air when heated.

B. Yttrium and oxygen.—Oxide of yttrium, Yttria, Y O. An insoluble, white earth; sp. gr. 4.840; it combines with acids to form crystallizable salts, which closely resemble those of glucina.

5. Erbium. Symb. E.

A. Exists as an oxide, associated with yttrium; very rare.

B. Erbium and oxygen. The oxide is insoluble in water; combines with acids (very little examined).

6. Terbium. Symb. Tb.

A. Exists as an oxide with yttrium; very rare.

B. Terbium and oxygen. A white, insoluble, earthy substance, forming pink salts with acids (very little examined).

7. **Thorium.** Symb. Th. Eq. 60. (59.59.)

A. Exists as an oxide in a very rare Swedish mineral, Thorite; obtained by the action of potassium on chloride of thorium; a malleable, gray metal, not acted on by water; burning in the air when heated.

B. Thorium and oxygen.—Oxide of thorium, thoria, ThO. A white, tasteless earth; combines with acids to form salts.

Sect. III.—Metals which tend more to form Acid Oxides than Basic Oxides.

1. **Tellurium.** Symb. Te. Eq. 64. (64.2.)

A. Exists native as a metal; both pure and alloyed with other metals; very rare; a brilliant, silvery-looking metal; sp. gr. 6.2; fuses at about 650°; crystallizes; and in many respects resembles sulphur and selenium.

B. Tellurium and oxygen. No protoxide, $Te\,O$, is known—A binoxide, $Te\,O_2$, or tellurous acid, is formed by burning the metal in air; a white, crystalline solid; fusible at a red heat; it is soluble in nitric acid and the alkalies, with which it combines to form tellurites. By the action of chlorine on tellurite of potassa, the acid is converted into telluric acid, $Te\,O_3$, which may be crystallized with $3\,H\,O$. A lemon-yellow solid, insoluble in cold water, soluble in boiling water; it combines with bases to form tellurates.

C. Tellurium and chlorine, iodine, &c. Formed direct.

D. Tellurium and hydrogen.—Telluretted hydrogen, $Te\,H$. Formed when an alloy of tin and tellurium is acted on by muriatic acid; a colorless gas; sp. gr. 4.489; soluble in water, and resembling sulphuretted hydrogen.

E. Tellurium and sulphur.—Sulphurets of tellurium. There are two sulphurets, obtained by action on tellurous and telluric acid by sulphuretted hydrogen; they consist of $Te S_2$ and $Te S_3$.

2. ARSENIC. Symb. As. Eq. 75.

A. Exists native pure; as an oxide; sulphuret; and combined with other metals. Obtained by distilling its oxides with charcoal or black flux; a brilliant, steel-gray, brittle, crystalline solid; sp. gr. 5.8; when heated does not fuse, but rises rapidly in vapor below redness; its vapor has a peculiar penetrating odor resembling that of garlic.

B. Arsenic and oxygen.—Oxides of arsenic. There are three oxides of arsenic, not one of which possesses basic properties. A protoxide, As O, is formed when arsenic is exposed to damp air.—When the metal is heated in the air it burns and forms white arsenic, or arsenious acid, $As O_3$; it is produced largely in the "roasting" of arsenical ores, and occurs native in various places. A white solid; sp. gr. about 3.6, which exists dimorphous, crystallized, and amorphous; it is soluble in water, highly poisonous, and combines with bases to form arsenites; the arsenite of potassa is employed in medicine.—When arsenious acid, or arsenic, is distilled with nitric acid, it is converted into arsenic acid, $As O_5$. A white, deliquescent, uncrystallizable solid; fusible at a red heat, and decomposed by a higher heat; sp. gr. about 3.7; soluble in water; highly poisonous, as well as all its salts. The arseniates resemble the phosphates. The arseniate of lime, 2 Ca O, H O, $As O_5$, may be formed direct; it occurs native as Pharmacolite with 5 H O, and as Haidingerite with 2 H O.

C. Arsenic and chlorine.—Chloride of arsenic. The metal burns in chlorine, and forms a dense oily liquid, which does not congeal at 0°; boils at 270°; and consists of $As Cl_3$. No chloride, $As Cl_5$, analogous to arsenic acid is known.

D. Arsenic and hydrogen.—Arseniuretted hydrogen, $As H_3$. Formed when nascent hydrogen comes in contact with arsenic, or any of its compounds; a colorless, transparent gas; sp. gr. 2.695; slightly soluble in water, but may be collected over water; becomes liquid at — 40° ; it has a nauseous odor, and is virulently poisonous, even in small quantities.

E. Arsenic and sulphur.—Sulphurets of arsenic. The protosulphuret, or Realgar, As S_2, exists native; and may readily be formed direct. A crystalline, transparent, red solid; sp. gr. 3.45; fuses and sublimes when heated; in the air it burns; it is used in the composition of white fire, and other pyrotechnical mixtures.—When sulphuretted hydrogen is passed through a solution of arsenious acid in dilute hydrochloric acid, Orpiment or sesquisulphuret of arsenic is formed, As S_3; a yellow, fusible, crystalline substance which may be sublimed in close vessels, but burns if heated in the open air; sp. gr. 3.45; it dissolves in solutions of the caustic alkalies; and is sometimes used as a dye; it is the basis of the pigment called King's yellow; this sulphuret is also found native. When sulphuretted hydrogen is passed through a strong solution of arsenic acid, a persulphuret is formed, As S_5; a yellow, fusible, volatile, combustible solid. These three sulphurets are sulphur acids, combining with other sulphurets to form sulphur salts.

F. Tests.—In consequence of the highly poisonous nature of arsenic, it is often of great importance to detect its presence, as in cases of poisoning, &c. Its volatility, when heated, the peculiar garlic-like odor of its vapor, its combination with hydrogen from which it is deposited in the metallic state on burning, and the formation of a yellow sulphuret soluble in ammonia, but insoluble in nitric acid, are excellent tests of arsenic and its compounds.

3. Chromium. Symb. Cr. Eq. 28. (28.15.)

A. Exists as an oxide in chromite of iron; in Shetland, France, India, Siberia, North America. Obtained with difficulty by intensely igniting its oxide with charcoal. A brilliant, white metal; sp. gr. 5.9; almost infusible.

B. Chromium and oxygen.—Oxides of chromium. There are five. A protoxide, Cr O, obtained by acting on the protochloride by potassa; it rapidly absorbs oxygen from the air, passing into a deutoxide, $Cr_3\ O_4$.—A sesquioxide is obtained by heating the chromate of mercury to redness, or by heating bichromate of potassa with charcoal; a green, crystalline solid, which occurs native associated with oxide of iron.—A binoxide, $Cr\ O_2$, is formed by igniting the nitrate.—Peroxide of chromium, or chromic acid, $Cr\ O_3$, is obtained by heating oxide of chromium with nitre, which forms a chromate of potassa, and this may be decomposed by sulphuric acid; chromic acid is a scarlet, crystalline solid, decomposed by heat, which converts it into sesquioxide; very soluble in water. Chromic acid appears to be the coloring matter of the Ruby; and both it and the green oxide are used in coloring porcelain and glass. Chromate of potassa is a yellow, crystalline salt, $K\ O,\ Cr\ O_3$; obtained by the action of nitre on the native chromite of iron; it is soluble in water, fusible at a red heat; sp. gr. 2.65; and is converted by sulphuric acid into a bichromate, $K\ O,\ 2\ Cr\ O_3$; an orange or red crystalline salt; used in calico printing, and as a source of the chromate of lead and other compounds used as pigments. Most of the chromates are yellow or red crystalline salts.

i. Sesquioxide of chromium and acids form salts; the soluble ones are mostly crystallizable; they are chiefly of a green color.

C. Chromium and chlorine.—Chloride of chromium. There are two chlorides—Cr Cl, and $Cr_2\ Cl_3$—formed by passing dry chlorine over a heated mixture of sesquioxide of chromium and charcoal; the sesquichloride may likewise be obtained by evaporating the hydrochloric solution of the sesquioxide; it is a pink, fusible, volatile solid, but forms a green solution in water.

D. The brilliant colors of the chromium compounds supply abundant and excellent tests of its presence.

4. Vanadium. Symb. V. Eq. 68. (68.55.)

A. Exists as an oxide, with other metals; exceedingly rare; Mexico, Great Britain, Sweden. Obtained by heating its chloride in dry ammonia; a white, silvery, brittle metal, not acted on by air or water; burns when heated in the air.

B. Vanadium and oxygen. There are three oxides: a protoxide, $V O$, obtained by passing dry hydrogen over heated vanadic acid; a binoxide, vanadous acid, $V O_2$, obtained by heating a mixture of the protoxide with vanadic acid—it is salifiable, most of its salts being blue. Vanadic acid, $V O_3$, is obtained by heating the vanadiate of ammonia, a salt obtained from the native vanadiate of lead by the action of nitric acid and ammonia.

5. Molybdenum. Symb. Mo. Eq. 48. (47.88.)

A. Exists as a sulphuret; rare; Norway, Great Britain, France, America. Obtained by passing hydrogen over ignited molybdic acid. A whitish gray, rather brittle, almost infusible metal; sp. gr. about 8.61.

B. Molybdenum and oxygen. There are three oxides: a protoxide, $Mo O$; a deutoxide, $Mo O_2$; and molybdic acid, $Mo O_3$; all three of these oxides are salifiable. Molybdic acid also unites to bases, and forms molybdates.

6. Tungsten. Wolframium. Symb. W. Eq. 100.

A. Exists as an oxide with other metals, as in Wolfram, &c.; rare; France, Great Britain, Germany, North America. Obtained by intensely heating a mixture of tungstic acid and charcoal; an iron-gray, hard, brittle, almost infusible solid; sp. gr. 17.4; oxidizes when heated in the air.

B. Tungsten and oxygen. There are two oxides: a binoxide, $W\ O_2$; and $W\ O_3$, or tungstic acid. The oxide is black; sp. gr. 12.1; is not salifiable, and heated in the air absorbs oxygen, and passes into tungstic acid; which may also be obtained from tungstate of lime, by boiling it in nitric acid; it is a yellowish-white powder; sp. gr. 6.0, insoluble in water. The tungstate of lime occurs native.

C. Tungsten and chlorine. Tungsten burns in chlorine, and forms a red, crystalline chloride, $W\ Cl_2$. When the sulphuret of tungsten is heated in chlorine it forms a terchloride, $W\ Cl_3$.

7. Antimony. Stibium. Symb. Sb. Eq. 129. (129.03.)

A. Exists native as a metal, and as a sulphuret, associated with other metals; obtained from the sulphuret by roasting, and subsequent reduction with charcoal. A brilliant, silvery-white, crystalline, brittle metal; sp. gr. 6.712; fuses at about 800°.

B. Antimony and oxygen. There are four oxides: a suboxide, Sb O;—a protoxide, $Sb\ O_3$; formed by strongly heating the metal in the air; a white, fusible, volatile, crystalline solid, which occurs native, and is salifiable, combining with acids;—a deutoxide, antimonious acid, $Sb\ O_4$, obtained by igniting the peroxide; it is white; sp. gr. 6.69; is infusible, and combines with bases;—a peroxide, antimonic acid, $Sb\ O_5$; formed by boiling the powdered metal in nitric acid; a pale yellow, tasteless solid; sp. gr. 6.25; infusible; loses oxygen at a red heat; combines with bases to form antimoniates; that of potassa is a soluble, crystalline salt, $K\ O, Sb\ O_5$; the antimoniate of soda is crystalline, and almost insoluble in water.

C. Salts of antimony. The protoxide forms salts with acids; the most important of these is the tartrate of potassa and antimony, or emetic tartar, $K\ O,\ Sb\ O_3,\ C_8\ H_4\ O_{10} + 2\ H\ O$. Formed by boiling oxide of antimony with bitartrate of potassa; a white, soluble, slightly efflorescent salt used in medicine. Other preparations of antimony are also used in pharmacy. James's powder consists of oxide of antimony and phosphate of lime; it varies in composition, and is a less certain preparation than emetic tartar.

D. Antimony and chlorine.—Chlorides of antimony. The metal burns in chlorine, and forms a soft solid sesquichloride, $Sb\ Cl_3$, the "butter of antimony." It may be obtained by distilling the powdered metal with perchloride of mercury, or dissolving it in hydrochloric acid and evaporating; it is decomposed by water; sometimes used to bronze gun barrels.—If exposed to excess of chlorine, a volatile liquid perchloride is formed, $Sb\ Cl_5$. This compound is also decomposed by water, yielding a white oxichloride, "Algaroth's powder," $Sb\ Cl_3 + 5\ Sb\ O_3$.

E. Antimony and hydrogen. They combine to form a colorless, gaseous compound, $Sb\ H_3$; when burnt, metallic antimony is deposited.

F. Antimony and sulphur.—Sesquisulphuret of antimony, $Sb\ S_3$. Occurs native, in Borneo, Hungary, France, England; the common ore of the metal; it may be formed direct; an iron-gray, brittle, fusible, crystalline solid; sp. gr. 4.66. It is used as a substitute for plumbago in the manufacture of inferior pencils.—There is also a persulphuret $Sb\ S_5$; and a curious oxisulphuret, termed the golden sulphuret; the composition of which does not appear to be definite.

G. Tests.—the soluble salts of antimony are all decomposed by zinc, which precipitates the metal; sulphuretted hydrogen throws down a brown-red or orange precipitate; heated with charcoal, most of the antimonial salts yield metallic antimony.

8. Columbium. Tantalium. Symb. Ta. Eq. 185. (184.9.)

A. Exists as an oxide combined with yttria in some very rare Swedish minerals. Obtained by acting on the potassiofluoride of columbium by potassium; a black or dark gray solid, having little metallic lustre; sp. gr. about 6.0; heated in the air it burns.

B. Columbium and oxygen. There are two, Ta O_2, and Ta O_3, or columbic acid, which combines with bases. The columbate of yttria occurs native as yttrotantalite.

9. **Pelopium.** Symb. Pl.

A. Exists as an oxide associated with columbium; exceedingly scarce (little examined).

10. **Niobium.** Symb. Nb.

A. Exists as an oxide with the preceding (little examined).

11. **Titanium.** Symb. Ti. Eq. 24. (24.29.)

A. Exists as an oxide; scarce. Obtained as a metal from the slags of iron furnaces, in which titaniferous ores have been smelted. Obtained by passing dry ammonia over ignited ammoniochloride of titanium; a brilliant, very hard, copper colored, crystalline solid; sp. gr. 5.3; apparently infusible.

B. Titanium and oxygen.—A protoxide, Ti O, is formed when the vapor of potassium is passed over ignited peroxide of titanium;—when heated in the air, it absorbs oxygen and passes into titanic acid, Ti O_2, which occurs native, nearly pure in Rutilite, Anatase, &c.; several of the titanates occur native.

C. Titanium and chlorine.—Chloride of titanium, Ti Cl_2. Formed direct; a dense, transparent, colorless liquid, which boils at about 210°; fumes in the air, and is decomposed by water.

Sect. IV.—Metals which tend more to form Basic Oxides than Acid Oxides.

1. **Gold.** Symb. Au. Eq. 200.

A. Exists native as a metal, pure, and associated with other metals; a brilliant, yellow, malleable, ductile solid; sp. gr. 19.2; transparent in thin leaves; fusible at a full, red heat, and crystallizing partially if cooled slowly; not acted on by any ordinary agents.

B. Gold and oxygen. There are two definite oxides: Au O, obtained by decomposing the protochloride by potassa; a dark violet solid;—and peroxide or auric acid, $Au\ O_3$, obtained by decomposing the perchloride by magnesia; a black, insoluble compound; both oxides are easily decomposed by heat, and the peroxide is also by light. When protochloride of tin is added to a solution of the perchloride of gold, a fine purple precipitate is formed which consists of peroxide of tin, protoxide of gold, and water; it is called purple of Cassius, and is used in porcelain painting. Peroxide of gold combines with ammonia to form aurate of ammonia, or fulminating gold; a greenish brown solid; it is very easily decomposed, and explodes violently when heated to 290°; it appears to consist of $2\ N\ H_3,\ Au\ O_3$.

C. Gold and chlorine.—Chlorides of gold. When gold is acted on by chlorine, or by nitromuriatic acid (aqua regia), which contains free chlorine in the nascent state, it is dissolved, and a perchloride of gold formed, $Au\ Cl_3$, a red, deliquescent, crystalline compound, soluble in water, ether, and alcohol; it is decomposed by light, heat, organic matter, and most metals; its etherial solution is used to gild steel, &c. When heated, it loses chlorine, and becomes a pale yellow protochloride, Au Cl, which is converted by water into perchloride and metallic gold. The perchloride of gold combines with other chlorides to form double salts, termed auroperchlorides; that of potassium, $K\ Cl,\ Au\ Cl_3 + 5\ H\ O$, is a yellow, slightly efflorescent, crystalline salt; that of sodium is also crystalline, and contains only 4 H O.

D. Gold and iodine. Gold and iodine do not combine direct, but when iodide of potassium is added to protochloride of gold, a greenish yellow, insoluble, crystalline protiodide is formed, Au I; it is very easily decomposed by heat, light, &c. When perchloride of gold is added to a solution of iodide of potassium, an insoluble, dark green periodide is formed, $Au\ I_3$;

it is easily decomposed, and, like the perchloride, forms double salts.

E. Gold and bromine. The aqueous solution of bromine dissolves gold, and forms a perbromide, $Au\ Br_3$; a dark red, crystalline salt, which resembles the chloride, and forms double salts.

F. Gold and sulphur. A persulphuret, $Au\ S_3$, is formed when sulphuretted hydrogen is passed through a solution of perchloride of gold; a black, insoluble powder, decomposed by heat. It combines with other sulphurets; with the alkaline hydrosulphurets it forms soluble salts.

G. Gold and cyanogen. There are two cyanides of gold, $Au\ C_2\ N$, and $Au\ 3\ C_2\ N$; they both form double compounds with other cyanides. The aurocyanides of potassium and sodium are easily formed; they are soluble salts, and are used for gilding metallic and other surfaces by electro-deposition.

H. Alloys.—Gold is generally alloyed with one-twelfth of pure copper; this forms standard gold, and is harder, though not more oxidizable, than the pure metal. In France one-tenth of copper is employed.

I. Tests.—The assay of gold is generally performed by cupellation; it is fused with lead, and kept for some time fused in a porous vessel made of bone-ash; the lead, and with it the other metals, except silver, are oxidized, and form a slag, which is absorbed by the bone earth; finally, the gold and silver thus left are separated by digestion in nitric acid, or "parting;" this dissolves the silver, and leaves the gold untouched. All the compounds of gold are reduced to the metallic state by a red heat.

2. Platinum. Symb. Pt. Eq. 99. (98.68.)

A. Exists native as a metal, associated with other metals; a white, malleable, ductile metal, capable of being welded or joined at a white heat; sp. gr. 21.53; fusible at a bright white heat; resembles gold in its general chemical relations; not acted on by ordinary agents. Finely divided platinum, which may be prepared in various ways, possesses remarkable catalytic powers, causing a number of chemical changes to take place merely by its presence; it possesses considerable power of absorbing and condensing gases.

B. Platinum and oxygen. Like gold, it has very little affinity for oxygen; there are, however, two definite oxides; a protoxide, Pt O, obtained by the action of potassa on the protochloride,—a peroxide, $Pt\ O_2$, formed by decomposing the persulphate by soda; both these oxides are salifiable. The protonitrate may be formed direct; it is soluble, of a deep brown color, and decomposed by heat, peroxide being left. The pernitrate may be also formed direct; it is dark brown, and soluble in water. The protosulphate is almost black; the persulphate is a black, deliquescent, very soluble salt.

C. Platinum and chlorine. When platinum is acted on by chlorine or nitromuriatic acid, a perchloride, $Pt\ Cl_2$, is formed; a dark red brown, crystalline solid, easily soluble in water, alcohol, and ether; heated to about 400°, it loses chlorine, and becomes a gray, insoluble protochloride, Pt Cl; this is decomposed entirely by a red heat. Both the chlorides of platinum combine with other chlorides to form double salts. The ammoniobichloride of platinum is formed when a solution of sal-ammoniac is added to a solution of perchloride of platinum; a yellow powder, almost insoluble in water, quite insoluble in solutions of sal-ammoniac and alcohol; it consists of $N\ H_3$, Cl H, $Pt\ Cl_2$, or $N\ H_4\ Cl$, $Pt\ Cl_2$, and when heated leaves finely divided or spongy platinum. The potassiobichloride, K Cl, $Pt\ Cl_2$, is also a difficultly soluble yellow salt; the formation of this compound affords a useful test for potassa; the corresponding sodium compound is soluble in water and alcohol, and crystallizes with 6 H O.

D. Platinum and iodine. They have no direct action on each other; but by the action of iodide of potassium on the solutions of the metal, three iodides, consisting of Pt I—Pt_2 I_3—and Pt I_2, may be obtained.

E. Platinum and sulphur. A protosulphuret, Pt S, may be formed direct, by heating finely divided platinum with sulphur. When perchloride of platinum is acted on by alkaline hydrosulphurets, a bisulphuret, Pt S_2, is formed.

F. Platinum and silicon. When platinum is long ignited in contact with charcoal and silica, it becomes brittle, and appears to contain about 3 per cent. of silicon; for this reason, the contact of platinum vessels with fuel at a high temperature should be avoided.

G. Uses.—Platinum is a metal of great importance to the chemist, in consequence of the high temperature it will sustain, and the great difficulty with which it is attacked by acids, &c.; the retorts or stills in which oil of vitriol is concentrated are made of platinum. Vessels of platinum are liable to be corroded by chlorine, phosphorus, the alkalies, and the oxides of easily reducible metals.

H. Alloys.—Many of the alloys of platinum are easily fusible, as with silver, arsenic, tin, zinc; it is for this reason that these metals and their oxides must not be heated in platinum vessels.

I. Tests.—The salts of platinum, like those of gold, are easily reduced to the metallic state by heat. The production of difficultly soluble potassio and ammonio-bichlorides is also very characteristic.

3. **Palladium.** Symb. Pd. Eq. 54. (53.3.)

A. Exists native as a metal, associated with platinum. A hard, dull white, malleable, ductile solid; sp. gr. 11.3; fusible at a full yellow heat; oxidizes when heated in the air.

B. Palladium and oxygen. The protoxide, Pd O, is obtained by heating the protonitrate to redness; it is black; if precipitated from a solution of the protonitrate by alkali, it falls as an orange hydrate.—By the action of potassa on the bichloride of palladium and potassium, a binoxide, $Pd\ O_2$, is obtained; it is black when anhydrous, but may be precipitated from its solution as a brown hydrate. The protoxide is salifiable; most of its salts are red, or brownish red; the nitrate may be formed direct by digesting the metal in nitric acid; the sulphate also may be formed in a similar manner.

C. Palladium and chlorine. When the metal is dissolved in nitromuriatic acid, a dark brown perchloride, $Pd\ Cl_2$, is formed; when heated, this leaves protochloride, $Pd\ Cl$, and by a further continuance of the heat, a subchloride is obtained. These chlorides, like those of gold and platinum, have great tendency to form double salts.

D. Palladium and sulphur, $Pd\ S$. Formed direct, by heating the metal with sulphur; a hard, white, fusible solid. It may also be formed by the action of sulphuretted hydrogen on salts of palladium; it then falls as a black powder; when roasted at a dull red heat, it loses sulphur, and is reduced to the metallic state.

E. Palladium and carbon. They appear to have very considerable affinity, and the metal, when retained at a red heat in contact with carbon, absorbs it and becomes very brittle.

F. Palladium and cyanogen.—Cyanide of palladium, $Pd\ N\ C_2$. Formed by adding cyanide of mercury to a neutral solution of any of the salts of palladium; it is decomposed by heat, and leaves the metal pure.

G. Uses.—Chiefly in the metallic state as a stopping material by dentists; pure, and alloyed with silver.

H. Alloys.—Palladium combines readily with tin, lead, arsenic, silver, &c., and forms fusible alloys. It forms a hard, white alloy with gold.

I. Tests.—Most metals precipitate palladium from its solutions; cyanide of mercury throws it down from neutral solutions; sulphuretted hydrogen precipitates it dark brown, the alkalies orange.

4. **Rhodium.** Symb. R. Eq. 52. (52.11.)

A. Exists native as a metal, associated with others in the ore of platinum. A very hard, white metal, very infusible; sp. gr. about 11.0; oxidizes when heated in the air.

B. Rhodium and oxygen.—Sesquioxide of rhodium, $R_2 O_3$. Is obtained by heating finely divided rhodium with nitre and caustic potassa; a dark gray or black powder; difficultly soluble in acids, but combining with them to form salts. The nitrate, formed direct, is a deep red, uncrystallizable salt; the sulphate, also formed direct, is brown red and uncrystallizable. Salts of a protoxide, R O, have also been formed, but the oxide itself has not been obtained separate; when the sesquisulphate is carefully heated, it leaves a protosulphate.

C. Rhodium and chlorine.—There are two chlorides, R Cl and $R_2 Cl_3$; they are red, soluble in water, uncrystallizable, and have great tendency to form double compounds with other chlorides. The rhodiochloride of potassium, 2 K Cl, $R_2 Cl_3$, is formed by acting on a mixture of rhodium and chloride of potassium by chlorine at a red heat; it crystallizes with 2 H O.

D. Alloys.—Several of the alloys of rhodium are remarkably hard, and have in consequence been employed in machinery, metallic pen making, &c.

E. Tests.—The salts of this metal are mostly red; they are easily reduced by heat, hydrogen, sulphuretted hydrogen, and other metals.

5. **Osmium.** Symb. Os. Eq. 100. (99.56.)

A. Exists native in the ore of platinum. A grayish black solid; sp. gr. about 10.0; infusible, not volatile, burning in the air when heated to redness.

B. Osmium and oxygen.—Oxides of osmium. There are four compounds of osmium and oxygen: a protoxide, Os O, obtained by the action of the alkalies on the protochloride; a sesquioxide, $Os_2 O_3$, which has only been obtained in combination; a binoxide, $Os O_2$, obtained by the action of carbonated alkalies on the bichloride; and a peroxide, osmic acid, $Os\ O_4$, formed by the combustion of the metal in oxygen, or boiling it in nitric acid. This is a crystalline, volatile, soluble, highly poisonous substance, resembling new bread in odor, and combining with bases to form salts.

C. Osmium and chlorine. There are four chlorides, analogous to the four oxides; they combine to form double chlorides.

6. **Iridium.** Symb. Ir. Eq. 99. (98.68.)

A. Exists native in the ore of platinum, as an alloy with osmium. A white, brittle metal; sp. gr. about 18.0; fusible at an intense white heat; oxidized by ignition in the air, but not by any acid.

B. Iridium and oxygen.—Oxides of iridium. There are four oxides: a protoxide, Ir O, obtained by acting on the protochloride by an alkali;—a sesquioxide, $Ir_2\ O_3$, formed when the metal is ignited with nitre;—a binoxide, $Ir\ O_2$, known only in combination with acids;—a peroxide, $Ir\ O_3$, obtained by the action of a solution of carbonate of potassa on the iridiochloride of potassium. Both the protoxide and the binoxide are salifiable; whether the peroxide is, has not been ascertained. The salts of iridium are mostly green, or greenish yellow, and uncrystallizable.

C. Iridium and chlorine. There are three chlorides: Ir Cl, $Ir_2\ Cl_3$, and $Ir\ Cl_2$. They form double chlorides; the bichloride has only been obtained in combination with chloride of potassium, &c.

7. **Ruthenium.** Symb. Rt. Eq. 52. (52.11.)

A. Exists native in the ore of platinum. A brittle, infusible metal; sp. gr. 8.6; which oxidizes when heated in the air.

B. Ruthenium and oxygen. Three oxides have been described: a protoxide, Rt O; obtained by heating the bichloride with carbonate of soda; it is black, and insoluble in acids;—a sesquioxide, $Rt_2 O_3$; a black solid, formed when the metal is heated in the air, or when alkalies are added to a solution of the sesquichloride, when the oxide is precipitated as an orange hydrate, soluble in acids;—a peroxide, $Rt O_2$; obtained by roasting the bisulphuret.

C. Ruthenium and chlorine. Chlorides of ruthenium have been formed; they unite to form double chlorides.

8. Silver. Argentum. Symb. Ag. Eq. 108.

A. Exists native pure; as a sulphuret; and combined with other metals, as lead, antimony, arsenic, &c. Obtained from its ores generally by amalgamation with mercury, and subsequent distillation of the latter metal. The native sulphurets are ground, roasted with common salt: the chloride thus formed decomposed by iron, and the reduced metal then collected as a fluid amalgam. Silver is separated from lead by cupellation, the lead being oxidized and converted into litharge, whilst the silver remains pure; its separation from lead is also facilitated by the fact, that an alloy of silver and lead is more fusible than pure lead. A brilliant, white, malleable, ductile metal; sp. gr. 10.5; fusible at a full, red heat; exhibits considerable tendency to crystallize; does not oxidize when heated in the air.

B. Silver and oxygen.—Oxides of silver. There are three: a suboxide, $Ag_2 O$, formed when silver salts of organic acids, as the citrate of silver, are heated in hydrogen;—a protoxide, $Ag O$, obtained by decomposing nitrate of silver by means of an alkali; a dark, olive-green solid; sp. gr. 7.2; reduced to the metallic state by heat, light, or hydrogen, and combines with acids to form salts; it communicates a fine yellow color to glass, and is used in porcelain painting. Protoxide of silver is soluble in a solution of ammonia, forming a clear, colorless solution; by the action of a very strong solution of ammonia on freshly precipitated oxide of silver, an insoluble black compound is formed, which is dangerously explosive, detonating with violence, even whilst moist; its exact nature is undetermined.—When solutions of silver are decomposed by electricity, a dark-colored, crystalline peroxide is formed, consisting of $Ag\ O_2$.

C. Protoxide of silver and acids.—Salts of silver.

i. Oxide of silver and chloric acid.—Chlorate of silver, $Ag\ O, Cl\ O_5$. Formed direct; a soluble, crystalline salt, which forms detonating mixtures with combustibles.

ii. Oxide of silver and iodic acid.—Iodate of silver, $Ag\ O, I\ O_5$. Formed by adding iodate of potassa to nitrate of silver; a white, insoluble salt, soluble in ammonia and in nitric acid.

iii. Oxide of silver and nitric acid.—Nitrate of silver, $Ag\ O, N\ O_5$. Formed by the action of dilute nitric acid on silver; the acid is decomposed, and the metal oxidized and dissolved. A white, crystalline salt; sp. gr. 4.35; easily soluble in water; fusible at a dull red heat; when fused, and cast into sticks, it constitutes the lunar caustic used in pharmacy. Exposed to light, particularly when organic matter is present, it blackens, and silver is reduced. Nitrate of silver is valuable to the chemist as a test for chlorine; it is used in medicine, and in the preparation of permanent marking inks and hair dyes.

iv. Oxide of silver and hyposulphurous acid.—Hyposulphite of silver, $Ag\ O, S_2\ O_2$. Formed when chloride of silver is acted on by hyposulphite of soda; a gray, soluble, intensely sweet compound; it forms a double salt with hyposulphite of soda.

v. Oxide of silver and sulphuric acid.—Sulphate of silver, $Ag\ O, S\ O_3$. Formed when silver is boiled in sulphuric acid containing a little nitric acid; or when solution of nitrate of silver and sulphite of soda are mixed; a white, crystalline, fusible, difficultly soluble salt; sp. gr. 5.34.

vi. Oxide of silver and phosphoric acid.—Phosphates of silver.

a. Phosphate of silver, tribasic phosphate of silver, 3 Ag O, P O_5. Is formed when tribasic phosphate of soda is added to a solution of nitrate of silver; it is a yellow powder, insoluble in water; fusible at a red heat; sp. gr. 7.3; soluble in nitric acid, and blackened by light.

b. Pyrophosphate of silver, bibasic phosphate, 2 Ag O, P O_5. Formed by the addition of pyrophosphate of soda to nitrate of silver. A white powder, insoluble in water; fusible at a red heat; sp. gr. 5.3; soluble in nitric acid, and decomposed by light.

c. Metaphosphate of silver, monobasic phosphate, Ag O, P O_5. A white, gelatinous salt, formed by the action of metaphosphate of soda on solution of nitrate of silver.

vii. Oxide of silver and carbonic acid.—Carbonate of silver, Ag O, C O_2. Obtained as a pale, yellowish white, insoluble compound, when carbonate of potassa is added to a solution of nitrate of silver.

viii. Oxide of silver and cyanic acid.—Cyanates of silver.

a. Cyanate of silver, Ag O, N C_2, O. Formed by adding cyanate of potassa to nitrate of silver; a white, difficultly soluble salt; soluble in ammonia.

b. Fulminate of silver, 2 Ag O, 2 N C_2, O_2. Obtained by the action of warm alcohol on nitrate of silver. A gray, crystalline powder, almost insoluble in cold water, but soluble at 212°; violently explosive when heated, or by the slightest friction. When acted on by a solution of potassa, it forms argentofulminic acid, Ag O, H O, 2 N C_2, O_2; the salts of this acid are mostly explosive; in them, the H O of the acid is replaced by one equivalent of the base.

c. Cyanurate of silver, 3 Ag O, 3 N C_2 O_3. Formed when nitrate of silver is added to a hot ammoniacal solution of cyanurate of ammonia; a white, insoluble salt, remarkably stable; not decomposed by light, nor by a heat of 600°. When nitrate of silver is added to a cold solution of cyanurate of potassa, a cyanurate of silver is also formed; it consists of 2 Ag O, H O, 3 N C_2, O_3.

ix. Oxide of silver and acetic acid.—Acetate of silver, Ag O, C_4 H_3 O_3. Formed when acetic acid is added to a solution of nitrate of silver; a white, crystalline salt, almost insoluble in water; nitrate of silver is sometimes useful as a test for acetic acid.

x. Oxide of silver and arsenious acid.—Arsenite of silver, Ag O, As O_3. Formed on adding a solution of an alkaline arsenite to nitrate of silver; a greenish-yellow, insoluble salt, soon darkened by exposure to light.

xi. Oxide of silver and arsenic acid.—Arseniate of silver, 3 Ag O, As O_5. Formed when arsenic acid, or a soluble arseniate, is added to nitrate of silver; a reddish-brown salt; insoluble in water, but soluble in nitric acid, and in ammonia.

xii. Oxide of silver and chromic acid.—Chromate of silver. Chromate of potash throws down a fine red, insoluble chromate of silver, Ag O, Cr O_3, when added to a solution of the nitrate; heat resolves it into oxide of chromium and metallic silver. Bichromate of potassa throws down from nitrate of silver a red, crystalline, slightly soluble bichromate of silver, Ag O, 2 Cr O_3.

D. Silver and chlorine.—Chloride of silver, Ag Cl.

Formed direct, or by double decomposition; silver and chlorine have a very powerful affinity for each other; it occurs native as horn silver; a white, fusible solid; sp. gr. 5.45 when fused; insoluble in water; volatile at a white heat; decomposed by light when moisture is present; it is not decomposed by heat alone, but is easily reduced when heated with an alkali; it is also decomposed by contact with zinc and dilute sulphuric acid. Chloride of silver is insoluble in nitric acid, but readily soluble in solutions of alkaline hyposulphites, and in solution of ammonia, in the latter case a kind of fulminating silver is sometimes formed. Nitrate of silver is a most valuable test for chlorine, and the strong affinity which the metal has for chlorine is often made use of in preparing various acids, as nearly all the salts of silver are completely decomposed by chlorine or muriatic acid. Chloride of silver forms an excellent photographic preparation; the remaining chloride not blackened by exposure to light, is easily removed by a solution of hyposulphite of soda.

E. Silver and iodine.—Iodide of silver, Ag I. Formed direct, or by double decomposition; a yellowish-white solid; insoluble in water and ammonia; very easily decomposed by light; it forms the basis of the film acted on in the daguerreotype and other photographic preparations; it occurs native in Mexico.

F. Silver and bromine.—Bromide of silver. Formed direct, or by double decomposition; an insoluble, crystalline, yellow, fusible compound; sp. gr. 6.3; rapidly blackened by exposure to light; it occurs native in Mexico, and also in France.

G. Silver and sulphur.—Sulphuret of silver, Ag S. Readily formed direct by heating sulphur and silver, or by the action of sulphuretted hydrogen on solutions of silver; it exists native, as one of the most abundant of the ores of silver, as vitreous silver ore, and associated with antimony as ruby silver ore; it constitutes the tarnish which forms on the surface of plate; a dark gray solid; sp. gr. about 7.0.

H. Silver and cyanogen. Cyanide of silver, $Ag\ N\ C_2$. Formed when hydrocyanic acid is added to a solution of nitrate of silver, or by double decomposition; a white solid; fusible below redness; insoluble in water; soluble in ammonia; decomposed by hydrochloric acid, and by sulphuretted hydrogen, hydrocyanic acid being set free.

I. Alloys. Silver combines with most metals to form alloys. With one-twelfth ($\frac{9}{120}$ths) of copper it forms standard silver as employed for coin; the proportion used in France is one-tenth. Silver, even in small quantity, renders the color of gold paler; standard gold formerly contained silver, but it is now only alloyed with copper.

K. Tests. The salts of silver are easily reduced by heat, light, hydrogen, and nearly all oxidizable metals; the soluble salts are precipitated by solutions containing chlorine. The assay of silver is generally conducted in the dry way, by cupellation or roasting in a porous earthen vessel with lead; the oxides of lead and other oxidizable metals form a fusible slag which is absorbed by the vessel, pure silver alone remaining.

9. Mercury. (Hydrargyrum.) Symb. Hg. Eq. 100. (100.07.)

A. Exists native in the metallic state, rare, as an amalgam of silver; chiefly as a sulphuret, pure, and with sulphuret of arsenic. Obtained from the sulphuret by distillation with iron, or with lime; a silvery white metal, fluid at common temperatures; sp. gr. 13.56; solid and crystalline at —39°; boils at 660°, forming a colorless vapor; rises in vapor at all temperatures above 40°; several valuable medicinal preparations consist of finely divided metallic mercury, such as blue pill, mercurial ointment, &c.; oxidizes slowly when heated in the air to 600°.

B. Mercury and oxygen.—Oxides of mercury. There are two salifiable oxides of this metal, a suboxide, Hg_2O, and a protoxide, $Hg\ O$. The sub or black oxide is obtained by acting on the protochloride by lime or potassa water; a blackish-brown powder, insoluble in water, gradually passing into protoxide and metallic mercury; this change is also more rapidly produced by light, and by heat; it is soluble in acids, and forms salts.—The red, or protoxide, is formed when heated mercury is exposed to the air; by heating the nitrate; or by the action of solution of potassa on the perchloride; a red, crystalline solid, slightly soluble in water (?); sp. gr. 11.07; it is gradually decomposed by light; a red heat resolves it into mercury and oxygen; it is soluble in acids, and forms salts with them. Most of the salts of this metal are highly poisonous; some of them form valuable medicines. The protoxide of mercury is soluble in ammonia, and forms with it a fulminating compound.

C. Suboxide of mercury and acids.

i. Suboxide of mercury and nitric acid. There are three salts, a nitrate, Hg_2 O, NO_5; formed when the metal is acted on by cold dilute nitric acid, a soluble, colorless salt, which crystallizes with 2 H O; and when heated leaves the red oxide.—When the crystals of this salt are left for some time in the mother liquor, a sesquinitrate is formed, 3 Hg_2 O, 2 N O_5; a white salt, soluble in a little water, but decomposed by a large quantity; it crystallizes with 3 HO.—When either of the preceding salts is thrown into a large quantity of water, a difficultly soluble, yellow subsalt is formed, 2 Hg_2 O, N O_5. When ammonia is added to the dilute solution of the neutral nitrate of the suboxide, a grayish-black ammonia-nitrate is formed, 3 Hg_2 O, N H_3, H O, N O_5.

ii. Suboxide of mercury and sulphuric acid, Hg_2 O, SO_3, thrown down when sulphate of soda is added to a solution of the nitrate of the suboxide; a white, difficultly soluble, crystalline salt; soluble in dilute nitric acid.

iii. Suboxide of mercury and carbonic acid, Hg_2 O,

CO_2. Formed by the reaction of solution of nitrate of the suboxide, and bicarbonate of potassa; a grayish-white insoluble salt.

iv. Suboxide of mercury and Fulminic acid, $2\ Hg_2\ O$, $2\ N\ C_2, O_2$; Fulminating mercury. Formed by the action of alcohol on a solution of nitrate of mercury under the influence of heat; a yellowish-gray crystalline powder, almost insoluble in cold, but soluble in boiling water; it explodes with considerable violence by friction, an electric spark, a temperature of 300°, contact of strong acids, &c.; it sometimes explodes whilst moist; much used in the manufacture of percussion caps; the force of the explosion is great, and its action very rapid, but the volume of gaseous matter evolved being small, it extends only a short distance, and has little propulsive force.

v. Suboxide of mercury and chromic acid, $4\ Hg_2\ O$, $3\ Cr\ O_3$. Formed when chromate of potassa is added to a solution of nitrate of the suboxide; an orange-red, insoluble salt, decomposed by heat, which leaves green oxide of chromium.

D. Protoxide of mercury and acids.

i. Protoxide of mercury and nitric acid. A solid, monobasic pernitrate of mercury has not been obtained; the solution of the protoxide in nitric acid, when evaporated, yields crystals of a white, soluble dipernitrate, $2\ Hg\ O,\ N\ O_5,\ 2\ H\ O$; this salt is decomposed by abundance of cold water, and then forms a difficultly soluble, pale yellow, tribasic nitrate, $3\ Hg\ O,\ N\ O_5,\ H\ O$; by the continued action of boiling water on this tribasic nitrate, it is converted into a reddish-brown, insoluble, sexbasic nitrate, $6\ Hg\ O,\ N\ O_5$.

ii. Protoxide of mercury and sulphuric acid. Persulphate of mercury, $Hg\ O,\ S\ O_3$. Formed by boiling mercury in sulphuric acid; sulphurous acid gas is evolved, and a white sulphate formed; by the action of water it is decomposed, and an almost insoluble, yellow subsulphate is formed consisting of $3 Hg O, S O_3$; and called turpeth mineral; it is a violent cathartic and emetic.

iii. Protoxide of mercury and carbonic acid, $4 Hg O, C O_2$. Formed when carbonate of potassa is added to a solution of the nitrate of the protoxide; a reddish-brown salt, insoluble in water, but soluble in water containing carbonic acid.

E. Mercury and chlorine. Chlorides of mercury.

i. Dichloride or subchloride of mercury; Calomel, $Hg_2 Cl$. May be formed direct, mercury being in excess; or by heating a mixture of the sulphate of the suboxide with common salt; or by adding a solution of common salt to a solution of nitrate of the suboxide. A white, or pale yellowish-white, crystalline solid; insoluble in water; sp. gr. 7.17; volatile at a heat a little below redness; the subchloride of mercury is occasionally found native; it is largely manufactured, and much used in medicine.

ii. Perchloride or bichloride of mercury.—Corrosive sublimate, $Hg Cl$. Formed direct, chlorine being in excess; the metal burns in chlorine when heated; obtained by heating a mixture of common salt with sulphate of the peroxide. A white, semi-transparent, imperfectly crystalline solid; sp. gr. 5.40; soluble in water, ether, and alcohol; when heated, it fuses, boils, and volatilizes; it has a peculiarly nauseous taste, and is a violently corrosive poison. Most forms of animal matter decompose corrosive sublimate, precipitating calomel; hence it is frequently used to render organic matter less prone to decay, and to prevent dry rot, &c. When heated in ammonia, bichloride of mercury absorbs it, forming a white, fusible, volatile compound, consisting of $2 Hg Cl, N H_3$. When solutions of sal-ammoniac and corrosive sublimate are mixed and evaporated, crystals are formed, of a compound of $Hg\,Cl, N\,H_3, Cl\,H$; this may be regarded as a double chloride of mercury and ammonium, $Hg\,Cl, N\,H_4\,Cl$. Corrosive sublimate also forms double chlorides with the other metallic chlorides.

F. Mercury and iodine. Iodides of mercury. When mercury and iodine are agitated together in alcohol, and when solutions of iodide of potassium and salts of the suboxide are mixed, a dingy green, nearly insoluble subiodide of mercury is formed, $Hg_2\,I$; sp. gr. 7.65; when heated, it fuses and sublimes; by the continued action of light or heat it is gradually decomposed. When iodide of potassium is added to an acid solution of the nitrate of mercury, a yellow, insoluble iodide is precipitated; it consists of $Hg_4\,I_3$; is volatile when heated, condensing in brilliant red crystals, which, however, become yellow again as they cool. When iodide of potassium is added to a solution of corrosive sublimate, an insoluble, red periodide is formed, $Hg\,I$; it is a brilliant red, crystalline compound, fusible and volatile when heated, and whilst hot, of a fine yellow color; sp. gr. 6.25; this iodide combines with other iodides to form double compounds, hence it dissolves in solutions of several iodides; when formed as above described, the precipitate redissolves in excess of iodide of potassium, or of corrosive sublimate.

G. Mercury and bromine.—Bromides of mercury. When bromine and mercury are agitated together, heat is evolved, and bromide of mercury formed, $Hg\,Br$. A difficultly soluble, crystalline, fusible, volatile solid; sp. gr. 5.9; when heated with excess of mercury, it forms a dibromide, $Hg_2\,Br$; a tasteless, yellowish-white, insoluble, fusible, volatile, crystalline solid; sp. gr. 7.3.

H. Mercury and nitrogen.—Nitruret of mercury, $Hg_3\,N$. Formed by passing ammonia over oxide of mercury heated to 300°; a brown powder, which explodes when heated, or struck with a hammer.

I. Mercury and amidogen.—Amide of mercury, $Hg\,N\,H_2$. This compound has not been hitherto isolated, but many double compounds containing it have been formed. When ammonia is added in slight excess to a solution of corrosive sublimate, a white, insoluble compound is formed, which is an amidichloride of mercury, $Hg\ H\ N_2$, $Hg\ Cl$; this salt has long been used in pharmacy, and forms a yet more complex compound with sal-ammoniac or chloride of ammonium, $Hg\ H\ N_2$, $Hg\ Cl + N\ H_3$, $H\ Cl$ (or $N\ H_4\ Cl$). There are also analogous amidoiodides, and amidobromides of mercury, as well as a very curious series of amidonitrates, consisting of $Hg\ N\ H_2 + 5\ Hg\ O$, $N\ O_5$.—$Hg\ N\ H_2 + 3\ Hg\ O$, $N\ O_5$.—$Hg\ N\ H_2 + 2\ Hg\ O$, $N\ O_5$, &c.

K. Mercury and sulphur.—Sulphurets of mercury. When nitrate of suboxide is added to a solution of bisulphuret of potassium, a black, insoluble sulphuret of mercury is formed, $Hg_2\ S$. It is decomposed by heat, and by long exposure to light. The protosulphuret, $Hg\ S$, is formed direct by heat, and by the action of sulphuretted hydrogen on the salts of the protoxide; it occurs native as cinnabar, the chief source of the metal; an insoluble, black or deep red solid, gradually decomposed by a dull red heat, or by distillation with iron, lime, &c.; when of a good color, and finely ground, it constitutes an excellent red pigment vermilion; it is volatile, and when freshly sublimed, is generally black, but becomes red when powdered; it occurs native, crystallized, and massive; pure, and associated with sulphuret of arsenic. The sulphuret of mercury forms double compounds with the chloride, iodide, and bromide of mercury.

L. Mercury and cyanogen.—Cyanide of mercury, $Hg\ N\ C_2$. Formed when oxide of mercury is boiled in water with Prussian blue, or is acted on by hydrocyanic acid; a white, crystalline solid; soluble in water; decomposed by heat, cyanogen being evolved, and by hydrochloric acid, hydrocyanic acid being formed;

it forms double salts with other cyanides, and also with chlorides, iodides, bromides, and various oxy-salts.

M. Alloys or amalgams. Mercury forms fluid or semi-fluid alloys with most of the metals; some of them are definite, and may be obtained crystallized.

N. Tests. The salts of mercury are all decomposed when heated with charcoal, metallic mercury being evolved; the metal is also precipitated from all solutions containing mercury by copper; the precipitation of the orange chromate, green, yellow, or red iodide, and black sulphuret, also afford useful tests of the metal.

10. COPPER. (Cuprum.) Symb. Cu. Eq. 32. (31.66.)

A. Exists native as a metal, as an oxide, chloride, sulphuret, and as a carbonate, phosphate, arseniate, &c.; obtained from the native sulphuret by roasting, which drives off the greater part of the sulphur and leaves an impure oxide; this is reduced by fusion with charcoal, and purified by repeated fusions, by which the impurities are oxidized and removed; a red, brilliant, malleable, ductile solid; sp. gr. 8.788 to 8.958; fusible at a yellow heat (about 1996°); boils and volatilizes at a white heat; occurs native crystallized, and may be obtained artificially in crystals; heated in the air, it oxidizes; in oxygen, it burns with a green flame.

B. Copper and oxygen.—Oxides of copper. There are two oxides, both of which unite to acids; a suboxide, $Cu_2 O$, which occurs native, as ruby copper and tile ore; it is obtained as a dark red solid, by heating the protoxide with metallic copper, or as an orange brown hydrate, by adding potassa to a hot solution of the subchloride; it is fusible at a red heat; heated in the air, it forms protoxide; it is soluble in ammonia, forming a colorless solution, which absorbs oxygen from the air on exposure to it, and forms then a deep blue solution of the protoxide; acids for the most part resolve it into protoxide and metallic copper, a few combine with it to form salts.—The protoxide, $Cu O$, is formed when copper is strongly heated in the air, or when the nitrate is heated to redness; a black solid, fusible at a bright red heat; sp. gr. 6.4; very hygrometric; insoluble in water; soluble in acids, soluble in caustic ammonia; decomposed even below redness by hydrogen and carbon, hence used as an oxidizing material in organic ultimate analysis; it may be obtained as a blue hydrate, by adding potassa to a solution of the sulphate; this hydrate is decomposed when moist, by a temperature of 212°; it is used in the arts, particularly in paper-staining, as a pigment, and is usually mixed with some insoluble salt of lime.

C. Protoxide of copper and acids.

i. Protoxide of copper and nitric acid.—Nitrate of copper, $Cu O, N O_5 + 3 H O$. Formed by the action of dilute nitric acid on copper; a blue, crystalline salt; very soluble in water and in alcohol; it may also be crystallized with $6 H O$; when the nitrate is heated to about 500°, it parts with water and acid, leaving a subnitrate $3 Cu O, N O_5, H O$. An insoluble, green powder, which may be also formed by the action of strong nitric acid on oxide of copper; this salt, when heated, yields the protoxide pure.

ii. Protoxide of copper and sulphuric acid.—Sulphate of copper; blue, or Roman vitriol, $Cu O, S O_3$. Formed direct, or by the exposure of roasted sulphuret of copper to air and moisture. White when anhydrous, but blue when dissolved in water, or as crystallized with $5 H O$; it may also be crystallized with $H O$. Used in the arts as a source of blue color; as a steep for seed-corn to prevent smut; as a preventive of dry rot; and in medicine. When sulphate of copper is boiled with excess of oxide, a green, insoluble disulphate is formed, $2 Cu O, S O_3$. There are also other sulphates of copper, containing $3 Cu O$,

SO_3, 3 HO—4 CuO, SO_3, 3 HO—5 CuO, SO_3 with 4 HO and 6 HO—8 CuO, SO_3 with 6 HO and 12 HO. Sulphate of copper forms double salts with other sulphates.

iii. Protoxide of copper and phosphoric acid.—Phosphate of copper. Several different phosphates may be obtained by precipitation, and others occur native; they are green, and insoluble in water; Trombolite, which occurs with the other ores of copper, consists of 2 CuO, PO_5, 4 HO, and Pseudomalachite, 5 CuO, PO_5, 5 HO.—Besides these, 6 CuO, PO_5, 3 HO, and 4 CuO, PO_5, 2 HO have been found native.

iv. Protoxide of copper and boracic acid.—Borate of copper, CuO, BO_3. Formed by double decomposition; a pale green, insoluble, fusible compound.

v. Protoxide of copper and silicic acid.—Silicate of copper, CuO, SiO_3, HO. May be formed by double decomposition with an alkaline silicate; it is a green, insoluble salt; it occurs native as Dioptase, or copper emerald. The silicate of the suboxide is a blue or purple vitreous compound, which, however, is very apt to acquire a green color from the absorption of oxygen.

vi. Protoxide of copper and carbonic acid.—Carbonate of copper. When carbonate of potassa is added to a cold solution of sulphate of copper, a blue sesquibasic carbonate is formed, 3 CuO, 2 CO_2, HO; if the solutions are hot, a green subcarbonate falls, 2 CuO, CO_2, HO; they are both prepared and used as pigments; the former constituting blue verditer, the latter mineral green; common blue verditer usually contains a considerable quantity of chalk; Malachite and mountain green are native carbonate of copper, and consist of 2 CuO, CO_2, HO; the sesquibasic carbonate also occurs native, constituting copper-azure, or mountain-blue; this salt may either be considered as 3 $Cu O, 2 C O_2, H O$; or as $Cu O, C O_2, H O + Cu O, C O_2, Cu O$; (a double hydrated carbonate, in which the water of the second equivalent is replaced by oxide of copper.)

vii. Protoxide of copper and acetic acid.—Acetates of $Cu O, C_4 H_3 O_3$. Formed direct by the action of vinegar on copper; known in commerce as crystallized verdigris; a bluish-green, soluble, crystalline salt; heated in the air, it burns; distilled, it yields a white, volatile, crystalline acetate of the suboxide, $Cu_2 O, C_4 H_3 O_3$. When copper is exposed to the fumes of vinegar, or plates of copper, covered with cloth steeped in vinegar, a sub or diacetate is formed; common blue verdigris, $2 Cu O, C_4 H_3 O_6 + 6 H O$; a greenish-blue, crystalline salt, partially soluble in, and partly decomposed by water; which forms two tribasic salts—the subsesquiacetate, $3 Cu O, 2 C_4 H_3 O_3 + 6 H O$, a pale blue, soluble, crystallizable salt; and an insoluble triacetate $3 Cu O, C_4 H_3 O_3 + 2 H O$. The different varieties of verdigris made, and met with in commerce, are mixtures of two or more acetates; they are also frequently largely adulterated.

viii. Protoxide of copper and arsenious acid.—Arsenite of copper, $Cu O, As O_3$. Formed by double decomposition with an alkaline arsenite; an apple-green, insoluble salt; much used as a pigment, and prepared under the names of Scheele's or Schweinfurth green.

ix. Protoxide of copper and arsenic acid.—Arseniate of copper, $3 Cu O, As O_5$; is formed as a pale greenish-blue, insoluble salt, by double decomposition with arseniate of potassa, &c. Various subarseniates occur native, as copper, mica, $8 Cu O, As O_5 + 12 H O$—Erinite, $5 Cu O, As O + 2 H O$, and also with $5 H O$ and $10 H O$,—Euchroite, $4 Cu O, As O_5$, &c.

x. Protoxide of copper and chromic acid.—Chromate

of copper, $Cu\,O, Cr\,O_3$. Formed when chromate of potassa is added to a solution of sulphate of copper; a buff-colored, nearly insoluble salt, having a sweetish taste, and soluble in ammonia.

D. Copper and chlorine.—Chlorides of copper.

i. Dichloride of copper, $2\,Cu\,Cl$. Formed direct when the metal is in excess, or by carefully evaporating a muriatic solution of the dioxide; a yellowish-brown, crystalline solid; fusible; resembling resin in appearance (sometimes called resin of copper); its vapor colors flame blue, green and red; insoluble in water, but soluble in hydrochloric acid; by the action of air and moisture, it forms a submuriate, or Brunswick green; this substance may also be formed by exposing copper, moistened with muriatic acid, to the air.

ii. Protochloride of copper, $Cu\,Cl$.—Formed direct, or by evaporating to dryness a muriatic solution of the protoxide; a yellowish-brown solid, soluble in water, and forming a green solution, which becomes blue when diluted; the green solution becomes temporarily green when heated to the boiling point; when partially decomposed by potassa, it yields a green, insoluble oxichloride, $3\,Cu\,O, Cu\,Cl + 4\,H\,O$; this compound occurs native in Peru as Atacamite. The chlorides of copper combine with other chlorides to form double salts.

E. Copper and iodine.—Diniodide of copper, $Cu_2\,I$. Formed direct, when copper leaf and iodine are heated together; white, and nearly insoluble in water; no protiodide is known; the diniodide combines with other iodides.

F. Copper and bromine.—Bromides of copper. When copper in excess is acted on by bromine, a dibromide is formed, $Cu_2\,Br$; it is soluble in water.—By dissolving protoxide in hydrobromic acid, a protobromide of copper is obtained; a gray, semi-metallic looking solid, like plumbago; soluble in water, and yielding deliquescent crystals, containing H O. They combine with other bromides.

G. Copper and hydrogen.—Hydruret of copper, Cu_2 H. Formed when a mixed solution of sulphate of copper and hypophosphorous acid is heated; it falls as a brown, insoluble powder; heated, it evolves hydrogen.

H. Copper and nitrogen.—Nitruret of copper, Cu_6 N. Formed by the action of ammonia on protoxide of copper, at a temperature of 480°; a green solid, which, heated in the air to 570°, ignites and gives off nitrogen.

I. Copper and sulphur.—Sulphurets of copper. When copper is heated with a small quantity of sulphur, it ignites, and a disulphuret is formed, Cu_2 S; a black, brittle solid, insoluble in water; sp. gr. 5.97. It occurs native as vitreous copper, one of the most important ores of the metal; it is found crystallized and massive; sp. gr. 5.70; has a shining lustre, and yields to the knife. The protosulphuret, Cu S, cannot be formed direct, but is thrown down when sulphuretted hydrogen acts on a protosalt of the metal; it also occurs native. The common copper glance, or pyrites, which is the chief ore of copper, consists of sulphuret of copper with iron; it also usually contains more or less arsenic.

K. Copper and carbon. They seem to combine; melted copper, which has been some time in contact with charcoal, becomes "overpoled;" it is brittle, fibrous, and appears to contain carbon.

L. Copper and cyanogen, a dicyanide, Cu_2 N C_2, may be formed by digesting the dioxide in hydrocyanic acid; it is white, insoluble in water, and combines with other cyanides; the cyanide is yellow, and insoluble in water.

M. Alloys. Many of these are of considerable importance in the arts, such as brass, bronze, bell-metal; copper renders gold and silver harder. The amalgam of copper is hard, but fusible at a low heat; it is sometimes used for taking casts.

N. The intense blue produced by the action of ammonia on copper compounds, and the easy precipitation of the metal from all its solutions by iron or zinc, are excellent tests.

11. URANIUM. Symb. U. Eq. 60.

A. Exists as an oxide in Pechblende and Uranite, very scarce; obtained by the decomposition of the chloride by potassium, when it is reduced and left as a black powder not acted on by air or water, but burning in the air when heated.

B. Uranium and oxygen.—Oxides of uranium. There are five: a suboxide, $U_4\ O_3$; a protoxide, $U\ O$; a black oxide, $U_4\ O_5$; a green oxide, $U_3\ O_4$; and a sesquioxide or uranic acid, $U_2\ O_3$. Of these, the protoxide, the green oxide, and the sesquioxide, are all salifiable; the sesquioxide is also an acid, combining with bases to form uraniates; the uraniate of ammonia, $N\ H_3$, $H\ O$, $2\ U_2\ O_3$, is a yellow, difficultly soluble salt; the uraniate of potassa, $K\ O$, $2\ U_2\ O_3$, as well as those of soda, lime, baryta, and magnesia, are also difficultly soluble yellow salts. Most of the salts of uranium are yellow or greenish-yellow. The peroxide is used to give a rich yellow color to glass; some of the colors might be used as pigments, were the metal more abundant.

12. BISMUTH. Symb. Bi. Eq. 213.

A. Exists native as a metal, as a sulphuret, and as an oxide; not very abundant. Obtained by reducing the oxide with charcoal or black flux; a brittle, brilliant, highly crystalline, reddish-white metal; sp. gr. 9.799; fusible at about 507°; volatile at a white heat.

B. Bismuth and oxygen.—Oxides of bismuth. There are three; a suboxide, formed when melted bismuth is long exposed to the air;—a protoxide, $Bi\ O_3$, obtained as a white powder by heating the subnitrate;—and a peroxide or bismuthic acid, Bi O_5, obtained by fusing the protoxide with caustic potassa; it is a red-brown powder, and combines with bases to form salts. The protoxide alone is salifiable.

C. Protoxide of bismuth and acids.

i. Protoxide of bismuth and nitric acid, Bi O_3, 3 N O_5 + 9 H O.—Formed when the metal is digested in dilute nitric acid; a colorless, crystalline salt, which dissolves in a little water, but is decomposed by the addition of much water, when a white, insoluble subnitrate is precipitated, which is called pearl-white; it consists of Bi O_3, N O_5; as formed, it is a hydrate, containing H O.

ii. Protoxide of bismuth and sulphuric acid.—Sulphate of bismuth, Bi O_3, 3 S O_3, is formed direct; a white salt, insoluble in water, but decomposed by it; a white, insoluble subsulphate being formed, Bi O_3, S O_3, together with an acid, soluble, crystallizable supersulphate.

iii. Protoxide of bismuth and carbonic acid.—Carbonate of bismuth, Bi O_3, C O_2; is thrown down by carbonated alkalies, as a white, insoluble compound; it contains 3 H O.

iv. Protoxide of bismuth and silicic acid, Bi O_3, 2 Si O_3; exists native, constituting bismuthite, bismuth blende.

D. Bismuth and chlorine.—Chloride of bismuth, Bi Cl_3; may be formed direct; the metal burns in chlorine; or bismuth may be distilled with perchloride of mercury, or the muriatic solution may be evaporated to dryness; a gray, deliquescent salt, fusible at 480°, and formerly called butter of bismuth; thrown into water, it is decomposed, and a white, insoluble, oxichloride precipitated, Bi Cl_3, 2 Bi O_3.

E. Bismuth and sulphur.—Sulphuret of bismuth, Bi S_3. Formed direct, and by the action of sulphuretted hydrogen in solutions of bismuth; it occurs native; a bluish, metallic-looking compound; sp. gr. 6.4.

F. Alloys.—With tin and lead, it forms fusible alloys. None of its alloys are of much practical value.

G. Tests.—the salts of bismuth are mostly decomposed by water; they are blackened by sulphuretted hydrogen, and are easily reduced to the metallic state when heated with charcoal.

13. Tin. Stannum. Symb. St. Eq. 59. (58.82.)

A. Exists native as an oxide; obtained by reduction of the oxide with charcoal or coal; a malleable, silvery white, brilliant metal, having a slight tinge of yellow; sp. gr. about 7.30; by friction, it exhales a peculiar odor; when bent, it gives out a remarkable crackling sound; it melts at 442°, and, if cooled slowly, crystallizes; it rapidly oxidizes, when heated, and burns brilliantly under a current of oxygen.

B. Tin and oxygen.—Oxides of tin. There are three oxides; a protoxide, Sn O, obtained by the action of potassa on a solution of the protochloride, when it falls as a hydrate, which must be dried out of contact of air; heated, it burns like tinder, and forms peroxide.—A sesquioxide, $Sn_2\,O_3$, is formed when protochloride of tin is acted on by sesquioxide of iron.—Peroxide of tin, Sn O_2, may be obtained by the action of nitric acid on the metal; it occurs native as a pale yellow, brown, or black crystalline solid; sp. gr. 7.0; the peroxide of tin has a strong affinity for coloring matter, and forms a valuable mordant; both the proto and the peroxide are salifiable. Peroxide of tin is used by jewelers as a polishing material, under the name of putty powder; fused with glass, it renders it white and opaque, constituting white enamel.

C. Protoxide of tin and acids.

i. Protoxide of tin and nitric acid, Sn O, N O_5, may be formed by digesting the metal, or hydrated protoxide in dilute nitric acid; a yellow, soluble, uncrystallizable salt; its solution absorbs oxygen rapidly from the air.

ii. Protoxide of tin and sulphuric acid, $Sn\ O,\ S\ O_3$; formed when the metal is digested in sulphuric acid; a white, soluble, crystalline salt.

D. Peroxide of tin and acids.

i. Peroxide of tin and nitric acid, $Sn\ O_2,\ N\ O_5$. Strong nitric acid acts violently on tin, forming peroxide, and nitrate of ammonia, a portion of water being decomposed; when the hydrated peroxide is digested in dilute nitric acid, a colorless, soluble, crystalline nitrate is formed.

ii. Peroxide of tin and sulphuric acid, $Sn\ O_2,\ S\ O_3$. Formed when the metal, or its hydrated peroxide, is boiled in sulphuric acid; an uncrystallizable, colorless salt. A solution of this sulphate forms the basis of "Bancroft's tin mordant."

E. Tin and chlorine.—Chlorides of tin. A protochloride of tin, Sn Cl, is formed by dissolving tin in hydrochloric acid, and carefully evaporating the solution; a gray, resinous-looking, fusible, volatile solid; decomposed by a large quantity of water, an oxichloride of tin being formed. Muriate of tin is commonly formed by digesting tin in muriatic acid; it is much used by dyers. Solution of protochloride of tin eagerly absorbs oxygen; it precipitates sulphur from sulphurous acid, and precipitates several metals in the metallic state; it throws down purple of Cassius from solution of gold; it forms a double salt with ammonia. Perchloride of tin is obtained by distilling tin with perchloride of mercury; a colorless, transparent liquid; it does not solidify at —20°, and boils at 250°; it fumes, when exposed to the air, and forms a crystalline hydrate with 5 H O. Muriate of tin is obtained by dissolving the metal in nitromuriatic acid; it is a valuable mordant; like the protochloride, it combines to form double salts.

F. Tin and iodine.—Iodides of tin. Protiodide of tin, Sn I, is formed by double decomposition; an orange-red, difficultly soluble, crystalline compound. Periodide of tin, $Sn\ I_2$, may be formed direct, with the assistance of heat; a yellow, fusible, crystalline solid, decomposed by water into hydriodic acid and peroxide of tin.

G. Tin and sulphur.—Sulphurets of tin. A protosulphuret, $Sn\ S$, may be formed direct with heat, or by the action of sulphuretted hydrogen on the protosalts; a black, fusible, crystalline solid, sp. gr. 5. When heated with sulphur, it forms a yellowish-gray sesquisulphuret, $Sn_2\ S_3$. When this is heated with sulphur and sal-ammoniac, a golden-yellow, crystalline bisulphuret is formed, $Sn\ S_2$; this substance is called mosaic gold, and is much used as a bronze powder. It occurs native, associated with sulphurets of iron and copper, as tin pyrites.

H. Alloys.—Tin combines with platinum readily, and forms a fusible alloy; they combine together at a red heat with considerable violence. Tin with copper ($Sn_{10}\ Cu_{90}$) forms gun metal, ($Sn_{20}\ Cu_{80}$) forms gong metal, ($Sn_{25}\ Cu_{75}$) forms bell metal, ($Sn_{34}\ Cu_{66}$) forms a good speculum metal, ($Sn_8\ Cu_{92}$) forms bronze; the metals combine with great facility, hence the process of tinning copper utensils for domestic use. Tin unites with mercury to form a semi-fluid amalgam; this is used for silvering mirrors. With iron it forms tin plate.

I. Tests.—The protosalts reduce the persalts of iron, copper, and mercury, &c., and precipitate solution of gold; all the soluble salts are precipitated by the alkalies and sulphuretted hydrogen; the metal is precipitated by zinc, and easily reduced by heat and charcoal.

14. Lead. Plumbum. Symb. Pb. Eq. 104. (103.56.)

A. Occurs native as a sulphuret, and as an oxide separate, and combined with various acids. Obtained from the sulphuret by roasting, which drives off most of the sulphur, and subsequent smelting or fusion with small coal. A bluish-gray, very soft metal, sp. gr. 11.445. It melts at 612°, and boils at a white heat, rapidly oxidizing; if cooled slowly, it crystallizes; it contracts considerably in solidifying, hence leaden castings, like bullets, have internal cavities.

B. Lead and oxygen. There are five oxides. A suboxide, $Pb_2\,O$, obtained when oxalate of lead is heated to about 570°, or when lead is long exposed to the air at about 650°, a dark gray, earthy-looking powder.—A protoxide, $Pb\,O$, obtained by heating the suboxide in the air, or by heating the carbonate or nitrate to dull redness; a yellow powder, fusible at a red heat, when it constitutes Litharge and Massicot; it possesses powerful fluxing properties, and is an important constituent of glass. A hydrated protoxide, $3\,Pb\,O + H\,O$, may be thrown down from salts of lead by alkalies, in excess of which, however, it dissolves; it is also formed when lead is placed in pure water, which is not the case if saline matter of any kind amounting to $\frac{1}{8000}$ is present; the hydrate is slightly soluble in pure water. The protoxide is salifiable.—The red oxide or minium, $Pb_3\,O_4$, is obtained by the continued action of heated air on the protoxide; a brilliant red powder; sp. gr. 8.6 to 9.0; not salifiable; used as a pigment; decomposed by heat, oxygen being evolved.—An orange-colored, crystalline sesquioxide, $Pb_2\,O_3$, is obtained by adding chloride of soda to a solution of oxide of lead in potassa.—A peroxide, $Pb\,O_2$, obtained by the action of nitric acid on minium; a dark brown powder, decomposed by a red heat, which converts it into protoxide; triturated with sulphur, it inflames; boiled in acids, oxygen is evolved, and a salt of the prot oxide formed; absorbs sulphurous acid. Peroxide of lead is a feeble acid (plumbic acid); it combines with potash and soda, to form difficultly soluble, crystallizable salts; most of its salts are insoluble, or decomposed by water; they are called plumbates.

C. Protoxide of lead and acids.

i. Protoxide of lead and nitric acid.—Nitrate of lead, $Pb\,O, N\,O_5$. Formed direct, or by digesting lead in dilute nitric acid; a white, crystalline, soluble salt; sp. gr. 4.0; decrepitates, when heated, and evolves nitric acid. When its solution is boiled with excess of oxide, a pearly white, crystalline, difficultly soluble dinitrate is formed, 2 Pb O, N O_5; this salt may be obtained dry, or crystallized with H O. When solution of the nitrate is digested with a slight excess of ammonia, a difficultly soluble tribasic nitrate is formed, 3 Pb O, N O_5 + 3 H O. By continued digestion in ammonia, this is resolved into a sexbasic nitrate, 6 Pb O, N O_5 + H O.

ii. Protoxide of lead and sulphuric acid.—Sulphate of lead, Pb O, S O_3. Formed direct, or by double decomposition with any salt of lead; a white, nearly insoluble powder, slightly soluble in solutions of the alkalies, their carbonates, and in concentrated sulphuric acid, from which it is precipitated on dilution. It occurs native, rare.

iii. Protoxide of lead and phosphoric acid.—Phosphates of lead.

a. Tribasic phosphate of lead, 3 Pb O, P O_5. Formed by adding solution of common tribasic phosphate of soda to acetate of lead; white, insoluble in water, fusible, decomposed by heat into metallic lead and bibasic phosphate; it occurs native, crystallized and associated with chloride of lead.

b. Bibasic phosphate of lead, 2 Pb O, P O_5. Formed by adding common phosphate of soda to a hot solution of chloride of lead; a white, fusible, crystallizable, insoluble salt.

c. Monobasic phosphate of lead, Pb O, P O_5. Formed by adding ammonia to a mixture of metaphosphoric acid and nitrate of lead. By the action of metaphosphate of soda in excess on chloride of lead a phosphate is formed consisting of 4 Pb O, 3 P O_5.

iv. Protoxide of lead and boracic acid.—Borate of lead,

$PbO, 2BO_3$. Formed by adding solution of borate of soda to nitrate of lead; a white, insoluble salt, which fuses into a very dense, transparent, pale yellow glass, used in the manufacture of optical glass.

v. Protoxide of lead and silicic acid.—Silicate of lead. May be formed by double decomposition with an alkaline silicate; white, insoluble in water, fusible at a red heat into a colorless glass. Exists in flint glass, and some of the glazes employed in making earthenware; mixes with the borate and other fusible compounds in all proportions. Oxide of lead is a powerful flux in consequence of the fusibility of its silicate, &c.

vi. Protoxide of lead and oxalic acid.—Oxalate of lead, PbO, C_2O_3. Formed by adding a solution of oxalic acid to nitrate of lead; a white, insoluble, crystalline compound; heated out of contact of air, it leaves a gray suboxide of lead. Oxide of lead also forms a basic oxalate, and several double oxalates; a crystallized oxalonitrate has been obtained, $PbO, C_2O_3 + PbO, NO_5 + 2HO$.

vii. Protoxide of lead and carbonic acid.—Carbonate of lead, white lead Ceruse, PbO, CO_2. Occurs native crystallized; formed by adding a solution of an alkaline carbonate to nitrate or acetate of lead; a white, crystalline powder, insoluble in water; sp. gr. about 6.5; decomposed by a dull red heat, which converts it into protoxide; manufactured on a large scale, chiefly by exposing lead to the fumes of vinegar and carbonic acid; used as a white pigment; when thus prepared, it is a white, uncrystalline, impalpable powder; it is highly poisonous, like most of the salts of lead; it dissolves with effervescence in nearly all acids, and is blackened by sulphuretted hydrogen. There is also a dicarbonate, $2PbO, CO_2 + HO$, formed by the action of pure water on lead; and a sesquibasic carbonate, $3\,Pb\,O, 2\,C\,O_2 + H\,O$.

viii. Protoxide of lead and acetic acid.—Acetate of lead.—Sugar of lead, $Pb\,O, C_4\,H_3\,O_3$. Formed direct, by digesting oxide of lead in vinegar or crude pyroligneous acid; a white, soluble salt, which forms crystals with $3\,H\,O$; sp. gr. 2.37; has a sweet, astringent taste, highly poisonous; fusible below redness, when it parts with acetic acid; it is partly decomposed by carbonic acid, which precipitates carbonate of lead. A sesquibasic acetate, $3\,Pb\,O, 2\,C_4\,H_3\,O_3$, is formed by the action of heat on the acetate; a diacetate, $2\,Pb\,O, C_4\,H_3\,O_3$, a difficultly soluble white salt, is formed by boiling oxide of lead in a solution of the acetate; a trisacetate, $3\,Pb\,O, C_4\,H_3\,O_3$, may also be formed in the same way; it is a difficultly soluble salt, used in pharmacy, constituting Goulard's lotion. The acetate is decomposed by carbonic acid, for which it constitutes an excellent test. By the action of excess of ammonia on the acetate, a white, difficultly soluble hexacetate is formed, $6\,Pb\,O, C_4\,H_3\,O_3$; it crystallizes with $3\,H\,O$. Acetate of lead is much used by dyers.

ix. Protoxide of lead and tartaric acid.—Tartrate of lead, $Pb\,O, C_8\,H_4\,O_{10}$. Formed by adding a solution of tartrate of potassa to nitrate or acetate of lead; a white, almost insoluble, crystalline powder; it then contains $4\,H\,O$; it may be rendered anhydrous, and when heated in a tube to dull redness, it leaves a mixture of finely divided metallic lead and charcoal, which is highly pyrophoric, igniting spontaneously on exposure to the air, and leaving protoxide of lead.

x. Protoxide of lead and arsenic acid.—Arseniate of lead. A white, insoluble bibasic arseniate of lead, $2\,Pb\,O, As\,O_5$, is formed when a solution of arsenic acid is added to nitrate of lead; it occurs native crystallized; a white, insoluble, fusible tribasic arseniate is formed when a solution of an alkaline bibasic arseniate is added to acetate of lead.

xi. Protoxide of lead and chromic acid.—Chromate of lead, $Pb\ O, Cr\ O_3$. A fine yellow or orange-yellow, insoluble salt; formed by adding a solution of an alkaline chromate to any solution of lead. It is much used as a pigment and dye material; in its yellow state it contains $H\ O$, but when thrown down orange from a hot solution it is anhydrous. Chromate of lead deepens in color when heated, it fuses at a red heat; it is easily decomposed by hydrogen or carbon even below redness, and not being hygroscopic is employed in place of oxide of copper in organic analysis; digested in potassa, it forms a scarlet, insoluble dichromate $2\ Pb\ O, Cr\ O_3$. Chromate of lead occurs native in red crystals; there is also a native sesquichromate, Melanochroit, $3\ Pb\ O, 2\ Cr\ O_3$.

D. Lead and chlorine.—Chloride of lead, $Pb\ Cl$. Formed direct by heating lead in chlorine; by digesting oxide of lead in muriatic acid; or by adding solution of chloride of sodium to nitrate or acetate of lead. A white, fusible, crystallizable solid; sp. gr. 5.13; difficultly soluble in cold water, more easily in boiling water; soluble in strong muriatic acid, almost insoluble in dilute muriatic acid; it occurs native. A fibrous yellow, crystalline dichloride, $Pb_2\ Cl$, has also been found native. An oxichloride of lead, $Pb\ Cl + Pb\ O$, may be formed by heating the chloride strongly in the air, or by fusing a mixture of the chloride and oxide; a native oxichloride, $Pb\ Cl + 2\ Pb\ O$, has been found. By the action of litharge on a solution of common salt, a hydrated oxichloride is formed; this, when dried and fused, constitutes a yellow pigment, Turner's or patent yellow.

E. Lead and iodine.—Iodide of lead, $Pb\ I$. Formed direct

with heat, or by double decomposition of iodide of potassium and nitrate of lead; a brilliant yellow, fusible, crystalline solid, sp. gr. about 6.0; difficultly soluble in cold water, more soluble in hot; it combines with ammonia to form a white iodoplumbate of ammonia, $NH_3 + PbI$, and combines with other iodides to form double salts. There are several distinct oxiodides; Pb I + Pb O, a greenish-yellow, fusible, insoluble compound is obtained when bibasic acetate of lead is decomposed by iodide of potassium;—Pb I + 2 Pb O, obtained in the same way with the tribasic acetate; and Pb I + 5 Pb O; obtained from the sexbasic acetate.

F. Lead and sulphur.—Sulphuret of lead, Pb S. Formed direct by fusion; heat being evolved; or by the action of sulphuretted hydrogen on any salt of lead; a gray, metallic-looking, brittle solid; sp. gr. 7.58. White lead and other compounds of the metal blacken, when exposed to the air, in consequence of the formation of sulphuret; the lead salts of organic acids are perfectly decomposed by exposure to sulphuretted hydrogen, a process often employed in their preparation; the insoluble sulphuret of lead and the excess of the gas being subsequently easily removed from the organic acid, which is thus obtained pure. Sulphuret of lead occurs native both massive and crystallized, as Galena, the chief ore of the metal; it often contains a small portion of silver; this may be separated from the lead by crystallization, when fused, or by cupellation; the presence of silver in lead is objectionable for several purposes, such as the manufacture of white lead, the color of which is injured by even a minute portion of any silver salt.

G. Lead and selenium.—Seleniuret of lead, Pb Se, may be formed direct by fusion; a gray metallic solid; occurs native; one of the chief sources of selenium.

H. Alloys.—Several of the alloys of lead are of considerable importance in the arts. With antimony in the proportions of 80 Pb + 20 Sb, it forms type metal;

a hard, yet fusible alloy, which takes clear sharp impressions when cast in a mould. With arsenic it forms an alloy which does not contract in solidifying like pure lead; $\frac{1}{100}$th part of arsenic is added to the lead used for shot making, it contributes to the spherical form of the shot. With silver it forms an alloy more fusible than pure lead, a fact which enables silver to be separated from lead with great facility. With copper, it forms a hard gray alloy, pot metal. With tin it forms several useful alloys; Pewter, which consists of 20 Pb + 80 St; common solder, 50 Pb + 50 St; and soft solder, which consists of 66 Pb + 34 St; these alloys are used conveniently in soldering lead, &c., being more easily fusible than it is. With tin and bismuth it forms fusible metal; 31 Pb + 19 St + 50 Bi; melts at 212°;—25 Pb + 25 St + 50 Bi; Rose's alloy melts at 200°; these alloys are chiefly interesting from the fact that, when cooling from 156° to 111°, they *expand*, below which point they contract like nearly all other solids.

I. Tests.—Sulphuretted hydrogen is a very delicate test for lead, the salts of which, in turn, supply an excellent test for sulphuretted hydrogen. The insoluble white sulphate, and the bright yellow iodide and chromate, are also excellent tests; the metal is precipitated from its solutions by zinc; and its compounds are easily reduced to the metallic state, when heated with charcoal.

15. Zinc. Symb. Zn. Eq. 32. (32.52.)

A. Exists native as an oxide, pure and combined with carbonic acid, and as a sulphuret. Obtained from the sulphuret by roasting which forms an impure oxide, which is decomposed by distillation with charcoal; the carbonate is reduced in the same manner. Zinc is sometimes called Spelter in commerce. A bluish-white, brittle, highly crystalline solid; sp. gr. about 6.8 to 7.2; tough, though brittle, at ordinary temperature, but malleable and ductile at about 300°, when it may be drawn into wire or rolled into plates; fusible at a dull red heat; volatile at a white heat; oxidizes in the air; burns when heated.

B. Zinc and oxygen. There appear to be three oxides of zinc;—a dioxide, $Zn_2\,O$; formed when the oxalate is carefully heated to redness out of contact of air;—a protoxide, Zn O, formed when the metal is heated in the air; it burns with a bright flame, and forms a light white, flocculent solid, "flowers of zinc," or philosophers' wool; a white solid; sp. gr. about 5.6; sometimes crystalline; becomes temporarily yellow, when heated; dissolves in alkaline solutions and in acids, and is salifiable; it has been employed as a white pigment; it may be obtained from its salts as a hydrate, with H O; both pulverulent and crystallized.—There is also a very easily decomposable peroxide of zinc.

C. Protoxide of zinc and acids.

i. Protoxide of zinc and nitric acid, $Zn\,O, N\,O_5 + 6\,H\,O$. Formed by the action of dilute nitric acid on zinc; a white, crystallizable, deliquescent, soluble salt. Two subnitrates have been formed; $4\,Zn\,O, N\,O_5 + 2\,H\,O$, and $8\,Zn\,O, N\,O_5 + 2\,H\,O$.

ii. Protoxide of zinc and sulphuric acid.—Sulphate of zinc; white vitriol, $Zn\,O, S\,O_3$. Formed by the action of dilute sulphuric acid on zinc; a white soluble salt, which crystallizes with 7 H O; and may be also crystallized with 6 H O and with 2 H O; it occurs native with sulphuret of zinc. There are also three subsulphates—$2\,Zn\,O, S\,O_3$—$4\,Zn\,O, S\,O_3$; and $8\,Zn\,O, S\,O_3 + 2\,H\,O$. The sulphate of zinc combines with other sulphates, to form double salts.

iii. Protoxide of zinc and carbonic acid.—Carbonate of zinc, $Zn\,O, C\,O_2$. Formed direct, or by double decomposition of sulphate of zinc and an alkaline carbonate; white, and insoluble in water. It occurs native as Calamine, one of the chief ores of the metal. The native carbonates of zinc vary in composition, and are colored with different substances; they occur both crystallized and massive, anhydrous and hydrated; some of them are subsalts. Carbonate of zinc forms double salts with other carbonates.

iv. Protoxide of zinc and fulminic acid.—Fulminate of zinc, $2\ Zn\ O, 2\ N\ C_2\ O$. Formed when zinc filings are shaken with fulminate of mercury and water; an insoluble, crystalline salt: it explodes on contact with sulphuric acid, by friction, and by a heat of 350°. One equivalent of the $2\ Zn\ O$ in this salt may be replaced by other bases, constituting a series of salts termed zincofulminates; thus the zincofulminate of potassa consists of $Zn\ O, K\ O, 2\ N\ C_2\ O$, &c.

v. Protoxide of zinc and silicic acid.—Silicate of zinc, $2\ Zn\ O, Si\ O_3 + H\ O$, may be formed by double decomposition. It occurs native, constituting electric calamine.

D. Zinc and chlorine.—Chloride of zinc, $Zn\ Cl$. Formed direct by heating zinc in chlorine, or by evaporating a muriatic solution of the metal; a white, semi-transparent, fusible, volatile, soluble, very deliquescent solid; it crystallizes as a muriate with $H\ O$. It is used in pharmacy, and as a preservative against dry rot for cordage, timber, &c. There are several oxichlorides, and the chloride combines with ammonia, and other chlorides to form double salts.

E. Zinc and iodine.—Iodide of zinc, $Zn\ I$. Formed by gently heating a mixture of zinc filings, iodine and water; a white, fusible, soluble compound; it forms double iodides.

F. Zinc and sulphur.—Sulphuret of zinc, $Zn\ S$. Formed by igniting a mixture of oxide of zinc and sulphur, or obtained hydrated by the action of alkaline hydrosulphurets on solutions of zinc. Sulphuret of zinc occurs abundant native as Blende, or black jack—both massive and crystallized, red, brown, and black;

by roasting, and exposure to air and moisture, it is gradually converted into sulphate of zinc.

G. Zinc and carbon.—Carburet of zinc, Zn C. Obtained as a black powder by distilling the cyanide.

H. Zinc and cyanogen.—Cyanide of zinc, Zn N C_2. Formed by the action of solutions of cyanide of potassium and sulphate of zinc; an insoluble, white powder.

I. Alloys.—The most important of the alloys of zinc are those it forms with copper. Brass or Dutch metal consists of from 64 Cu + 36 Zn to 93 Cu + 7 Zn. Brass for hammering usually contains 70 Cu + 30 Zn; frequently a small quantity of lead and tin is added. Brass solder consists of 50 Cu + 50 Zn. Princes' metal, pinchbeck, tombac, and similor, are alloys of copper and zinc in different proportions. Collins' and Muntz's yellow metal for sheathing, which admits of being rolled hot, consists of 60 Cu + 40 Zn. A kind of brass beaten out into thin leaves, and subsequently reduced to powder, constitutes one of the best bronze powders; it is tinged of various colors by superficial oxidation. In consequence of its great oxidizability, zinc is frequently used to protect less oxidizable metals from rust, or the action of sea water.

K. Tests.—The compounds of zinc are mostly colorless, and soluble in water; the alkalies throw down white, Zn O + H O, soluble in excess. The borate, carbonate, and phosphate are insoluble in water, but soluble in both acids and alkalies. Alkaline hydrosulphurets precipitate it pale yellowish white; zinc is easily reduced with charcoal and flux, and the color of the flame with which it burns is very characteristic.

16. Cadmium. Symb. Cd. Eq. 56. (55.74.)

A. Exists as a sulphuret, associated with the ores of zinc; obtained by distilling its oxide with charcoal; or precipitated from its solutions by zinc; a tin-white brilliant metal, rather harder than tin, crackles like it when bent; sp. gr. about 8.60; melts below 600, and distils at a dull red heat; oxidizes, when heated in the air.

B. Cadmium and oxygen.—Oxide of cadmium, Cd O. Obtained by precipitating the nitrate by an alkaline carbonate, and igniting the precipitate; an orange-brown, insoluble solid; sp. gr. 8.18; neither fusible nor volatile; very easily reduced, when charcoal is present; it may also be precipitated as a white hydrate, and is then soluble in ammonia, but insoluble in the fixed alkalies; it is salifiable.

C. Oxide of cadmium and acids.

i. Oxide of cadmium and nitric acid.—Nitrate of cadmium, $Cd\ O,\ N\ O_5 + 4\ H\ O$. Obtained by acting on the metal by dilute nitric acid; a white, crystalline, soluble, deliquescent salt.

ii. Oxide of cadmium and sulphuric acid.—Sulphate of cadmium, $Cd\ O,\ S\ O_3 + 4\ H\ O$. Formed by acting on the metal by dilute sulphuric acid; a colorless, soluble, crystallizable salt.

iii. Oxide of cadmium and carbonic acid.—Carbonate of cadmium, $Cd\ O,\ C\ O_2$. Obtained by the action of solution of an alkaline carbonate on sulphate, or nitrate of cadmium; a white, insoluble salt; decomposed at a red heat; the hydrated oxide of cadmium absorbs carbonic acid, when exposed to the air.

D. Cadmium and chlorine.—Chloride of Cadmium, Cd Cl. Formed by dissolving the metal in hydrochloric acid, and evaporating to dryness; a colorless, transparent, pearly-looking, fusible solid; volatile at a high heat; soluble in water, and forming efflorescent, hydrated crystals; chloride of cadmium absorbs ammonia, and forms a double salt with sal ammoniac; (this may be represented as $Cd\ Cl + N\ H_4\ Cl$.)

E. Cadmium and sulphur.—Sulphuret of cadmium, Cd S. Formed direct by heating sulphur with cadmium, or by the action of sulphuretted hydrogen on the salts of cadmium; a bright yellow solid; sp. gr. 4.90; neither fusible nor volatile at a red heat; fusible at a white heat; insoluble in ammonia; soluble in nitric acid; its color is very brilliant, and it forms a valuable pigment; it occurs native crystallized, as Greenockite.

F. The alloys of cadmium are not practically used.

G. Tests.—The metal is precipitated from its solutions by zinc; the yellow sulphuret is an excellent test; it is distinguished from sulphuret of arsenic by its insolubility in ammonia, and want of fusibility.

17. Nickel. Symb. Ni. Eq. 28.

A. Exists native as a metal, associated with arsenic and sulphur; obtained by igniting the oxalate, and subsequently reducing the oxide, thus formed, with charcoal at a white heat; a brilliant white metal; sp. gr. varies from 8.27 to 9.00; fusible at a full white heat; magnetic when cold, but loses this property above 630°; it is ductile and malleable; not oxidizable in the air at common temperatures, but oxidizes when heated.

B. Nickel and oxygen.—Oxides of nickel. There are two. A protoxide, Ni O, obtained as an apple green hydrate, by adding potassa to sulphate of nickel; it is soluble in ammonia, forming a fine sapphire-blue solution, purple by transmitted light; it is soluble in acids, and salifiable; heated, it becomes anhydrous, and is gray.—A sesquioxide, or peroxide, $Ni_2 O_3$, is obtained by heating the nitrate or carbonate; a black, insoluble powder, unsalifiable; sp. gr. 4.846; ignited, it evolves oxygen, and forms Ni O; it may be obtained as a black hydrate, by passing chlorine through the hydrated protoxide, suspended in water; it contains 3 H O.

C. Protoxide of nickel and acids.

i. Protoxide of nickel and nitric acid.—Nitrate of nickel, $Ni\,O, N\,O_5 + 5\,H\,O$. Formed when dilute nitric acid acts on the metal; a bright, green, soluble, crystallizable salt; deliquescent in damp, but efflorescent in dry air; heated, it loses water, and subsequently a portion of its acid, forming a basic salt.

ii. Protoxide of nickel and sulphuric acid.—Sulphate of nickel, $Ni\,O, S\,O_3$. Formed by dissolving the oxide or carbonate in dilute sulphuric acid, or by digesting the metal in sulphuric acid with the addition of a little nitric acid; an apple-green soluble salt, which crystallizes with 7 H O, forming transparent, emerald green, efflorescent crystals; it combines with other sulphates to form double sulphates.

iii. Protoxide of nickel and carbonic acid, $Ni\,O, C\,O_2$. Formed by the action of a solution of an alkaline carbonate on sulphate or nitrate of nickel; it is precipitated as an apple-green, insoluble powder, containing 3 H O; it dissolves in water containing excess of carbonic acid.

D. Nickel and chlorine.—Chloride of nickel, Ni Cl. The metal burns, when heated in chlorine; the chloride may also be formed by carefully heating the muriate to dryness; a greenish-yellow solid, volatile at a red heat, soluble in water, forming a green solution, from which it crystallizes with 9 H O; chloride of nickel absorbs $3\,N\,H_3$, and forms a bulky white powder.

E. Nickel and sulphur.—Sulphurets of nickel. When nickel filings and sulphur are heated together, they combine with ignition, and form a yellow sulphuret Ni S; it occurs native as hair pyrites. By the action of sulphuretted hydrogen on neutral solutions of nickel, a black, hydrated sulphuret is precipitated; this is not the case if the solution be acid. By heating a mixture of carbonate of nickel with carbonate of potassa and sulphur, a bisulphuret of nickel is formed, $Ni\,S_2$; and by passing hydrogen over the ignited sulphate, a subsulphuret is obtained, $Ni_2\,S$.

F. Alloys.—Nickel forms alloys of some importance with copper and zinc; Electrum consists of about 51 Cu + 26 Ni + 23 Zn; Tutenag of 46 Cu + 37 Zn + 17 Ni; German silver, 52 Cu + 23 Zn + 26 Ni; the inferior sorts contain less nickel; sometimes a small quantity, about 3 per cent., of lead is added; the Pakfong, or white copper of the Chinese, is a similar alloy. Nickel also exists in meteoric iron. An arseniuret of nickel may be formed by heating the arseniate with charcoal; native arseniurets exist as Kupfernickel, As Ni_2; and nickel pyrites, As Ni; the chief ores of the metal. —As Ni_3, is also found native crystallized.

G. Tests.—The salts of nickel are mostly green, the blue solution formed by its oxide in ammonia is very characteristic, and is easily distinguished from copper; if acid be present the metal is not precipitated by sulphuretted hydrogen; the alkalies and their carbonates precipitate it apple-green.

18. Cobalt. Symb. Co. Eq. 30. (29.52.)

A. Exists native as an oxide, and as a sulphuret; associated with nickel and arsenic. Obtained by reduction of the oxalate with charcoal, or by the action of hydrogen on the ignited oxide. A reddish-gray, brittle metal, very difficult of fusion; sp. gr. 7.834.

B. Cobalt and oxygen.—Oxides of cobalt; there are two oxides—a protoxide, Co O, formed by the action of potassa on a solution of the nitrate; or by heating the carbonate out of contact of air; a greenish-gray solid, soluble in acids, and salifiable—and a sesqui or peroxide, Co_2 O_3, formed when chlorine is passed through the hydrated protoxide suspended in water; it is black, may be obtained as a black hydrate, containing 3 H O; is not salifiable; and is sometimes found native. When the protoxide is heated in the air it forms an oxide intermediate between these two, consisting of Co_3 O_4; it appears, however, to be a mixture.

C. Oxide of cobalt and acids.

i. Oxide of cobalt and nitric acid.—Nitrate of cobalt, Co O, N O_5. Obtained by dissolving the metal or carbonate in dilute nitric acid; a red, crystalline, deliquescent salt, which heated leaves peroxide; soluble in water, and crystallizes with 6 H O; its solution furnishes a sympathetic ink, invisible when cold, red when warmed. Nitrate of cobalt is not precipitated by ammonia, which forms a double salt.

ii. Oxide of cobalt and sulphuric acid.—Sulphate of cobalt, Co O, S O_3. A difficultly crystallizable salt, formed direct or from the carbonate; when anhydrous it is blue, but it forms a pink solution, and crystallizes with 6 H O. When sulphate of cobalt and sulphate of zinc are mixed, and the solution precipitated by an alkaline carbonate, a green compound of the oxides of zinc and cobalt is obtained; it is sometimes used as a pigment, and called Rinmann's green. Oxide of cobalt also forms a bisulphate, and a series of double sulphates.

iii. Oxide of cobalt and phosphoric acid.—Phosphate of cobalt, Co O, P O_5. Formed by the action of solutions of phosphate of soda and chloride of cobalt: a lilac-colored, insoluble salt; when mixed with pure and freshly precipitated alumina, and ignited, it forms a beautiful blue pigment, Thenard's blue.

iv. Oxide of cobalt and silicic acid.—Silicate of cobalt. A blue silicate may be formed by double decomposition with a soluble alkaline silicate. When oxide of cobalt is fused with glass an intensely blue-colored glass is formed; this powdered constitutes smalt; the impure oxide of cobalt obtained by roasting the pounded ore mixed with a certain quantity of pounded flint is termed Zaffre; fused with alkali it forms smalt; a valuable blue pigment, used in coloring paper, china, glass, &c.

v. Oxide of cobalt and carbonic acid.—Carbonate of cobalt, Co O, C O_2. Formed easily by double decomposition with an alkaline carbonate; an insoluble pink powder, soluble in excess of carbonic acid, and decomposed by heat.

vi. Oxide of cobalt and arsenic acid.—Arseniate of cobalt, $3\ Co\ O, As\ O_5$. May be formed by double decomposition with an alkaline arseniate; a pink, insoluble salt; it occurs native with $3\ H\ O$, as peach-blossom cobalt ore.

D. Cobalt and chlorine.—Chloride of cobalt, $Co\ Cl$. Formed direct, the metal burns when heated in chlorine; or by carefully evaporating the muriate out of contact of air; a blue solid, readily soluble in water, forming a pink solution; and crystallizing of a pink color with $5\ H\ O$. A dilute solution forms a good sympathetic ink, invisible when cold, blue when dried by heating the paper.

E. Cobalt and sulphur.—Sulphurets of cobalt. When sulphate of cobalt is heated with charcoal, a protosulphuret is formed, $Co\ S$; it is yellowish-gray, fusible at a red heat, and decomposed by acids; by the action of sulphuretted hydrogen on the protosalts of cobalt, it is thrown down as a black hydrate.—When sulphate of cobalt is decomposed by sulphuretted hydrogen at a red heat, a dark-gray sesquisulphuret is formed, $Co_2\ S_3$; it occurs native.—When the oxide of cobalt is heated with excess of sulphur, a black bisulphuret is formed, $Co\ S_2$.

F. Cobalt and cyanogen.—Cyanide of cobalt, $Co, N\ C_2$. Formed when hydrocyanic acid is added to a solution of acetate of cobalt, when it falls as a brown hydrate; it is blue when anhydrous.—There is also another cyanide called cobaltocyanogen, $Co_2\ 6\ N\ C_2$; this combines with H_3 to form cobaltocyanic acid, which, acting on various metallic oxides, forms a class of cobaltocyanides. The cobaltocyanide of potassium, $Co_2\ 6\ N\ C_2, 3\ K\ O$, is a yellow, soluble, crystallizable salt; the cobaltocyanide of cobalt is pink, and insoluble.

G. Alloys.—None of these are of any importance; the chief use of the metal is in the manufacture of zaffre and smalt. The arseniuret of cobalt, $Co_2\ As_3$, occurs native crystallized, one of the chief ores of the metal; in Speisskobalt it is associated with iron and nickel.

H. Tests.—Hydrosulphuret of ammonia precipitates cobalt black; the salts are mostly red; the fixed alkalies throw down a blue, hydrated oxide, which passes through green into dirty red, if boiled; cobalt is not precipitated by other metals; the blue tinge given to glass is a marked character.

19. Iron. Ferrum. Symb. Fe. Eq. 28.

A. Exists native in the metallic state, rare, as meteoric iron, &c.; as an oxide pure, and associated with clay; as a carbonate, as a sulphuret, and associated with many other substances; obtained by the reduction of its oxide with charcoal at a white heat; a gray, malleable, ductile solid; fusible at a bright white heat; capable of being welded or joined at a white heat; sp. gr. 7.77; attracted by the magnet; and capable of acquiring magnetism by induction at common temperature, but not at a red heat; oxidizes very slowly in dry air, but rusts rapidly in moist air; oxidizes rapidly when heated; burns in the air at a white heat; or in oxygen, throwing off brilliant scintillations. The rusting of iron is prevented by contact with zinc, and increased by contact with less oxidizable metals, such as copper.

B. Iron and oxygen.—Oxides of iron. There are four oxides. The protoxide, $Fe\,O$, is obtained as a white hydrate when pure protosalts of iron are decomposed by an alkali, the contact of air being prevented; the water being driven off by heat, it remains as a greenish, friable powder, not magnetic, very greedy of oxygen, absorbing it eagerly from the air, and evolving heat; it also absorbs carbonic acid; it is soluble in acids, and forms salts with them, which are all very prone to oxidize.—The sesquioxide, $Fe_2\,O_3$, obtained by igniting the sulphate; a red or yellowish brown powder, soluble in acids, and salifiable; it may be thrown down from its salts as a bulky, red-brown hydrate; dried at 212°, it retains $3\,H\,O$; at 400°, it retains H O; and becomes anhydrous at about 550°: after ignition, it is very difficultly soluble in acids. This oxide is called Colcothar, and used by jewelers and others as a polishing material. The sesquioxide of iron occurs native, constituting Hæmatite, Specular iron ore, &c.; as a hydrate it occurs in black and brown hæmatite, ochre, umber; and associated with a variable quantity of clay, as Clay iron-stone.—The black or magnetic oxide, $Fe_3 O_4$, sometimes regarded as a compound of the two preceding oxides (Fe O, $Fe_2 O_3$), constitutes forge scales; is formed when iron is highly heated in the air, or by passing steam over ignited iron, or by boiling a mixture of the hydrated protoxide and sesquioxide in water; it is black and magnetic; it occurs native, constituting magnetic iron ore.—Ferric acid, Fe O_3, is formed when the sesquioxide is ignited with nitrate of potassa; a ferrate of potassa is then formed; by the decomposition of this, other ferrates may be obtained; ferric acid is very easily decomposed, its solution evolving oxygen, and depositing sesquioxide of iron.

C. Protoxide of iron and acids.

i. Protoxide of iron and nitric acid.—Protonitrate of iron, Fe O, N O_5. Formed by the action of dilute nitric acid on iron, or its protosulphuret (Iron sometimes assumes a passive condition, in which state it is not acted on by strong nitric acid; this appears to be an electric effect). The protonitrate is soluble in water; forms pale-green crystals, containing 7 H O; its solution rapidly absorbs oxygen, and, if boiled, becomes brown from the formation of peroxide.

ii. Protoxide of iron and sulphuric acid.—Protosulphate of iron, Fe O, S O_3. Formed by the action of dilute sulphuric acid on iron, or its protosulphuret; or by the action of moist air on the roasted bisulphuret; it is white when anhydrous, but green when crystallized with 7 H O, as in Copperas or common green vitriol; its aqueous solution is also green, and as well as the crystallized salt, rapidly becomes brown when exposed to the air from the formation of the peroxide; it is slightly efflorescent; heated to 238° it fuses, and parts with 6 H O; at 530° it loses the remaining H O, and becomes white and anhydrous; at a red heat it is resolved into sulphate of the peroxide, and sulphurous acid; and at a high red heat this is decomposed into anhydrous sulphuric acid, and peroxide of iron. Sulphate of iron may be crystallized with 2 H O, 3 H O, and 4 H O; it occurs native. Sulphate of iron forms double salts with other sulphates; thus with potassa it forms Fe O, K O, $2\ S\ O_3 + 6$ H O; with sulphate of alumina it forms an alum analogous to the corresponding potassa salt, called iron alum; it consists of $Fe\ O, S\ O_3 + Al_2\ O_3, 3\ S\ O_3 + 24$ H O. A salt of this sort occurs native in aluminous shales, termed plumose alum; it is used in the manufacture of potash alum (p. 116).

iii. Protoxide of iron and phosphoric acid.—Protophosphate of iron, 2 Fe O, H O, $P\ O_5$. Formed by adding a solution of common phosphate of soda to protosulphate of iron; white, fusible, insoluble; but soluble in most acids; it becomes blue by exposure to air from oxidation. A native protophosphate of iron occurs, consisting of 3 Fe O, $P\ O_5 +$ H O; it is massive, and crystallized, either white or blue.

iv. Protoxide of iron and carbonic acid.—Protocarbonate of iron, Fe O, $C\ O_2$. Formed by adding a solution of an alkaline carbonate to protosulphate of iron, when it falls as an insoluble white hydrate; by the action of air it speedily peroxidizes; it is soluble in water containing carbonic acid, and exists thus held in solution in many mineral springs. Protocarbonate of iron occurs abundantly native, as spathose iron ore, and associated with silica, alumina, &c., as clay iron-stone.

a. Reduction of iron ores. In countries where wood is abundant, the hæmatites and other pure oxides of iron are reduced direct by fusion with charcoal. The Norwegian, Swedish, and Russian iron is thus obtained; in England the clay iron-ore is the chief source of the metal; this contains from 40 to 80 per cent. of protocarbonate of iron. The ore is broken, and made into heaps with coal, which being set on fire is allowed to burn slowly so as to roast the ore; or it is roasted like limestone in close kilns; the roasted ore containing oxide of iron together with lime, magnesia, silica, alumina, &c., in variable proportion, is mixed with coke or raw coal, and lime, and fused in a blast-furnace urged either with hot or cold air; the carbonaceous matter reduces the iron, whilst the lime forms a fusible slag with the earthy matters of the ore, which protects the newly reduced metal from oxidation; the result of this process is pig or cast-iron, it contains carbon and earthy impurities; the process is continuous, fresh ore, &c., being from time to time thrown in above, whilst the reduced iron collects below, and is let off periodically; the slag continually flows off through an opening left on purpose. The proportions of the materials are varied according to the richness of the ore, the rich and poor ores being generally so assorted as to give an average yield of metal.

b. Manufacture of bar-iron.—Cast-iron goes through two processes to render it malleable. First, the refining, in which the pig-metal is fused in contact with charcoal, and secondly, puddling, in which the metal fused in an appropriate reverbatory furnace, termed a puddling furnace, is thoroughly stirred and agitated in contact with air and flame; it gradually becomes tough, and less fusible, when it is withdrawn from the furnace and reduced to shape under the tilt-hammer; the fused earthy impurities are thus squeezed out, and the metal becomes malleable; by continued hammering and rolling, it gradually becomes more free from impurities, and constitutes pure bar or wrought iron; the latter part of the process entirely depends on the welding property of iron.

v. Protoxide of iron and chromic acid. No chromate of the protoxide of iron exists; they decompose each other, and form oxide of chromium and peroxide of iron; a compound of protoxide of iron and oxide of chromium, however, occurs native, constituting the chief ore of chromium; it is termed chromite of iron, and sometimes, though improperly, chromate; it usually consists of $2\ Cr_2\ O_3 + Fe\ O$, or $3\ Cr_2\ O_3 + 2\ Fe\ O$.

D. Peroxide of iron and acids.

i. Peroxide of iron and nitric acid.—Pernitrate of iron, $2\ Fe_2\ O_3$, $3\ N\ O_5$. Formed by the action of hot diluted nitric acid on iron; a brown, deliquescent, crystalline salt, soluble in water, forming a brown solution, and crystallizing with $16\ H\ O$; it is decomposed by a red heat; carbonate of potassa throws down from its solution a hydrated peroxide, soluble in excess, and forming the alkaline tincture of iron of pharmacy. There is also a difficultly soluble, brown, gelatinous subpernitrate.

ii. Peroxide of iron and sulphuric acid.—Persulphate of iron, $Fe_2\ O_3$, $3\ S\ O_3$. Formed by boiling a solution of the protosulphate with a little nitric acid; a difficultly crystallizable, deliquescent, brown salt, white when anhydrous, insoluble in sulphuric acid, soluble in water; its dilute solution when heated deposits a dark orange, insoluble, basic sulphate, $3\ Fe_2\ O_3,\ S\ O_3 + 4\ H\ O$; monobasic—bibasic—sesquibasic—and bipersulphates may also be formed. These salts combine to form double sulphates.

iii. Peroxide of iron and phosphoric acid.—Perphosphate of iron, $2\ Fe_2\ O_3,\ 3\ P\ O_5$. An insoluble, white salt; formed by double decomposition with an alkaline phosphate and persulphate of iron; digested in an alkaline solution, it forms a brown, insoluble subperphosphate. The phosphate occurs native, combined with the protophosphate, constituting native "Prussian blue." The perphosphate also occurs native alone, hydrated.

iv. Peroxide of iron and oxalic acid.—Peroxalate of iron, $Fe_2\ O_3,\ 3\ C_2\ O_3$. Formed direct; a difficultly soluble, pale yellow, crystalline powder; soluble in oxalic acid; hence the use of oxalic acid in removing stains of iron or "iron mould" from linen, &c. There are also double oxalates with potassa, &c.

v. Peroxide of iron and carbonic acid. There is no percarbonate of iron.

vi. Peroxide of iron and acetic acid.—Peracetate of iron, $Fe_2\ O_3,\ 3\ C_4\ H_3\ O_3$. Formed by digesting iron clippings in acetic acid or vinegar; a reddish brown, soluble, uncrystallizable salt; much used by dyers as a mordant. Its solution deposits a yellow diacetate when rapidly evaporated.

vii. Peroxide of iron and gallic acid.—Pergallate of iron. When gallic acid is added to a solution of the black oxide of iron in an acid, a black, insoluble compound of the per and protogallate is formed; the gallate of the protoxide is soluble and colorless.

viii. Peroxide of iron and tannic acid.—Pertannate of iron, $Fe_2\ O_3,\ 3C_{16}\ H_8\ O_{12}$. Formed when excess of tannic acid is added to solution of persulphate of iron; gradually formed by oxidation when protosulphate is employed; a black insoluble salt; associated with a portion of pergallate it constitutes the coloring matter of common black ink, being held in suspension by gum or mucilage; common black ink may be removed from paper by oxalic acid; this is prevented in various ways; in the blue black inks sulphate of indigo is added.

ix. Peroxide of iron and arsenic acid.—Perarseniate of iron, $2\ Fe_2\ O_3, 3\ As\ O_5$. Obtained by the action of solutions of bibasic arseniate of soda and perchloride of iron; a white, insoluble, hydrated salt, containing $12\ H\ O$; red, when anhydrous. Several arseniates of iron occur native; as, the cubic crystallized arseniate, $Fe\ O, Fe_2\ O_3, As\ O_5 + 6\ H\ O$;—Skorodite, $2\ FeO, 2\ Fe_2\ O_3, 3\ As\ O_5 + 12\ H\ O$;—$2\ Fe_2\ O_3, As\ O_5$; is also found native; as well as a double hydrated persulphate and perarseniate, "pitch ore," $4\ Fe_2\ O_3, S\ O_3 + 2\ Fe_2\ O_3, 3\ As\ O_5 + 45\ H\ O$.

E. Iron and chlorine.—Chlorides of iron. There are two, corresponding to the two salifiable oxides. The protochloride, $Fe\ Cl$, is formed by passing hydrochloric acid over red-hot iron, or by evaporating the muriatic solution of the protosulphuret out of contact of air, and heating the residue to redness; a gray, fusible, crystalline solid, volatile at a high red heat; heated in the air it forms perchloride and peroxide of iron; it forms a green solution in water, and crystallizes in transparent, greenish-blue crystals, containing $4\ H\ O$. Protochloride of iron absorbs ammonia, and forms double chlorides. The sesquichloride or perchloride of iron, $Fe_2\ Cl_3$, is formed when iron is heated in chlorine; the metal burns, and forms a red brown, crystalline, volatile solid; it is very deliquescent, and easily soluble in water, forming a deep brown solution, from which it may be crystallized with $5\ H\ O$ or $12\ H\ O$. Sesquichloride of iron is soluble in alcohol, forming a valuable pharmaceutical preparation. The sesquichloride also absorbs ammonia, and forms double chlorides.

F. Iron and Iodine.—Protiodide of iron, Fe I, is formed when iron turnings are digested with iodine in water; a green solution is obtained, which yields it in green crystals, containing 5 H O, or by careful evaporation out of contact of air as a gray, fusible solid; its aqueous solution decomposes when kept, the metal peroxidizing, unless free iron is present. The periodide or sesquiodide, $Fe_2 I_3$, is formed by dissolving the hydrated peroxide in hydriodic acid.

G. Iron and nitrogen.—Nitruret of iron. When ammonia is passed over ignited iron, the latter absorbs about 10 per cent. of nitrogen, forming a white, brittle, magnetic solid; sp. gr. 5.0.

H. Iron and sulphur.—Sulphurets of iron. There are several. The protosulphuret, Fe S, is obtained by the action of sulphur or its vapor on white hot iron; a dark, bronze-colored, brittle, semi-crystalline solid, more fusible than iron; occurs native as magnetic pyrites. Coals free from sulphur must be used in the smelting of iron ores, or a portion of protosulphuret is formed, which seriously injures the quality of the pig-metal obtained.—The protosulphuret may be obtained as a black hydrate by the action of a solution of hydrosulphuret of ammonia on protosulphate of iron.—Bisulphuret of iron, $Fe S_2$, is formed by passing sulphuretted hydrogen over oxide of iron, heated nearly to redness; a brilliant yellow, metallic-looking solid; occurs abundantly native, both massive and crystallized, as iron pyrites; used as a source of sulphate of iron, and as a source of sulphur; exposed to the air it oxidizes, and, in any quantity, often produces so much heat as to cause ignition; beds of coal containing pyrites are not unfrequently thus set on fire.—A sesquisulphuret of iron, $Fe_2 S_3$, is obtained by passing sulphuretted hydrogen over peroxide of iron at 200°; or hydrated by the addition of the neutral persulphate to solution of an alkaline hydrosulphuret.—There is also a disulphuret, $Fe_2 S$, and a tetrasulphuret, $Fe_4 S$.—An arsenical sulphuret of iron occurs native, as arsenical pyrites, or Mispickel, $Fe As, Fe S_2$; a variety of this ore also contains a small portion of sulphuret of silver.

I. Iron and phosphorus.—Phosphurets of iron. Fe P has not been obtained; but a sesquiphosphuret, $Fe_2 P_3$, is obtained by passing phosphuretted hydrogen over the heated sulphuret of iron; and a phosphuret, $Fe_4 P$, is formed when the vapor of phosphorus is passed over red-hot iron. Cold short-iron appears to contain a portion of phosphuret, the presence of which is supposed to render it peculiarly brittle.

K. Iron and carbon.—Carburets of iron. No protocarburet has been obtained. A sesquicarburet, $Fe_2 C_3$, is formed by heating pure Prussian blue (sesquiferrocyanide of iron) to redness. A bicarburet, $Fe C_2$, is obtained in the same manner by igniting the ferrocyanide of ammonium; and by long-continued fusion of steel with charcoal, a compound of $Fe_4 C$, is obtained; this approaches very nearly to the composition of white cast-iron.—Cast-iron appears to be a subcarburet of iron; but, besides carbon, it also contains a small proportion of silicon, manganese, and sometimes phosphorus; in its conversion into bar-iron, the carbon is oxidized and burnt away; hence this process is facilitated by the addition of oxide of iron to the melted metal in the puddling furnace; sometimes, too, for the same purpose, peroxide of manganese is employed. Cast-iron is sometimes decarbonized by embedding it in powdered hæmatite, and heating it to redness; the carbon is thus abstracted from the cast-iron, and it becomes malleable and tough. When pure wrought-iron is embedded in charcoal powder, and retained for some hours in contact with it at a high temperature, it absorbs from $\frac{1}{90}$th to $\frac{1}{120}$th its weight of carbon, which appears to penetrate through the metal; it becomes steel: this process is termed

Cementation. Steel possesses the valuable property of temper; it may be rendered hard at pleasure, by heating it and suddenly quenching it in cold water; and hard steel, by carefully heating it, may be brought down to any required degree of hardness, elasticity, &c. Good steel appears always to contain a minute quantity of phosphorus. By careful management in the reduction of the native oxide of iron with charcoal, steel may be obtained direct from the ore; it is then termed native steel. In the process of cementation iron swells and blisters, hence crude steel is termed blistered steel; when drawn out and hammered, it constitutes tilted steel; when cut into pieces, welded, and again drawn out, it is shear steel; when fused, it forms cast steel.

L. Iron and cyanogen.—Cyanides of iron. There are three. The protocyanide of iron, $Fe\ N\ C_2$, is obtained as a yellow-gray powder when the ammoniocyanide is heated out of contact of air.—The sesquicyanide, $Fe_2\ 3\ N\ C_2$, has not been obtained separate, but only in combination with other substances.—A percyanide or ferrocyanogen, $Fe\ 3\ N\ C_2$, likewise is only known in combination; this very remarkable substance enters into a variety of combinations, giving rise to a series of interesting acids and salts. Ferrocyanogen combines with the metals to form ferrocyanides; it is bibasic, and forms two classes of compounds, those containing two equivalents of one metal, and those with single equivalents of two different metals.

i. Ferrocyanogen and hydrogen.—Ferrocyanic acid, $Fe\ 3\ N\ C_2,\ H_2$. Obtained by decomposing a solution of ferrocyanide of potassium by tartaric acid, or the ferrocyanide of barium by dilute sulphuric acid; a colorless, soluble, crystallizable, acid; inodorous, not poisonous, strongly acid; forms ferrocyanides of the metals when it acts on their oxides; with ammonia it forms a ferrocyanate (ferrocyanide of ammonium), $2\ N\ H_3,\ Fe\ 3\ N\ C_2,\ H_2$ or $2\ N\ H_4,\ Fe\ 3\ N\ C_2$,—a pale-yellow, soluble, crystallizable, transparent salt; the crystals contain 3 H O.

ii. Ferrocyanogen and potassium.—Ferrocyanide of potassium, ferroprussiate of potassa, K_2, Fe 3 N C_2. Formed when iron filings are digested in a solution of cyanide of potassium, or when Prussian blue is boiled in a solution of potassa; a yellow, transparent, easily soluble, crystallizable salt; sp. gr. 1.83; the crystals contain 3 H O; heated, it becomes white and anhydrous; heated in the air, it is resolved into cyanate of potassa and peroxide of iron. This salt is manufactured largely for dyers, &c., by igniting in a thick iron pot, animal matter, such as hoofs, horn, dried blood, &c., with iron clippings and carbonate of potassa, and subsequently lixiviating the residue, which by crystallization yields the salt pure; used as a source of the other metallic ferrocyanides, some of which, as Prussian blue, are valuable pigments; useful to the chemist as an excellent test of metals, most of which form colored, insoluble ferrocyanides. By the action of dilute sulphuric acid it yields a yellow, soluble, crystallizable ferrocyanide of potassium and iron, a salt in which one equivalent of potassium is replaced by one of iron, K Fe, Fe 3 N C_2. Similar compounds may be formed with calcium, barium, and other metals.

iii. Ferrocyanogen and sodium.—Ferrocyanide of sodium, Na_2, Fe 3 N C_2. Obtained by boiling Prussian blue with soda or its carbonate; a yellow, soluble salt, which forms efflorescent crystals with 12 H O.

iv. Ferrocyanogen and silver.—Ferrocyanide of silver. Ferrocyanide of potassium forms an insoluble, cream-colored precipitate in solutions of nitrate of silver.

v. Ferrocyanogen and mercury.—Ferrocyanide of potassium throws down an insoluble ferrocyanide of mercury in solutions of nitrate of mercury.

vi. Ferrocyanogen and copper.—Ferrocyanide of potassium throws down a deep red brown ferrocyanide of copper in solutions of copper; this is a delicate and useful test of the metal; it has been employed as a pigment and in calico printing.

vii. Ferrocyanogen and tin.—Ferrocyanide of potassium throws down a pale-yellow ferrocyanide in solutions of tin.

viii. Ferrocyanogen and lead.—Ferrocyanide of potassium forms a white, insoluble ferrocyanide in solutions of lead.

ix. Ferrocyanogen and zinc.—Ferrocyanide of potassium throws down a pale-yellowish white ferrocyanide in solutions of zinc.

x. Ferrocyanogen and iron.—Ferrocyanides of iron. Ferrocyanide of potassium forms a white precipitate in pure protosalts of iron, which, however, by oxidation rapidly becomes blue. This salt appears similar to, if not identical with, the ferrocyanide of potassium and iron already mentioned. With the salts of the sesquioxide, a deep blue, insoluble substance, Prussian blue, or sesqui-ferrocyanide of iron is formed; it consists of 4 equivalents of iron and 3 of ferrocyanogen, or Fe_4, $3 Fe 3 N C_2$; it is of a rich, intense blue, insoluble in water, very hygroscopic; concentrated sulphuric acid renders it white, but does not decompose it, as water restores its blue color; boiled with oxide of mercury and water it forms cyanide of mercury; alkalies decompose it, hence Prussian blue does not stand washing with soap; it is a valuable pigment, and of great use to the dyer, &c.; it forms an excellent blue ink with sulphate of indigo. The Prussian blue, as commonly prepared, always contains a portion of ferrocyanide of potassium and water.

LL. There is also another isomeric form of ferrocyanogen

termed Ferridcyanogen, the equivalent of which is double that of ferrocyanogen—Fe_2 6 N C_2; this substance forms a series of compounds distinct from, yet analogous to the preceding; it has not been isolated itself.

i. Ferridcyanogen and hydrogen.—Ferridcyanic acid, Fe_2 6 N C_2, H_3. Obtained by decomposing the ferridcyanide of lead by sulphuretted hydrogen; a soluble, crystallizable acid, which spontaneously decomposes when kept. It forms ferridcyanides by acting on the metallic oxides.

ii. Ferridcyanogen and potassium.—K_3, Fe_2 6 N C_2. Formed by the action of chlorine on a solution of the ferrocyanide of potassium; a ruby red, crystallizable, soluble salt; its solution furnishes precipitates of insoluble ferridcyanides with various metallic solutions, and constitutes a useful test. It precipitates silver orange, mercury yellow, copper and bismuth yellowish brown, tin white, zinc orange, &c.

iii. Ferridcyanogen and iron.—Ferridcyanide of iron, Fe_3, Fe_2 6 N C_2. No precipitate is caused in solutions of the pure persalts of iron by ferridcyanide of potassium, but in solutions of the protosalts it forms an insoluble, deep blue precipitate, which is manufactured as a pigment, Turnbull's blue.

M. Iron and sulphocyanogen.—Sulphocyanides of iron. A pale-green, soluble, protosulphocyanide, Fe, N C_2 S_2, is obtained by digesting iron in hydrosulphocyanic acid. A persulphocyanide, 2 Fe, 3 N C_2 S_2; obtained by dissolving the peroxide in hydrosulphocyanic acid; a deep red, deliquescent salt; the color of this compound is so intense, that sulphocyanide of potassium forms an excellent test for peroxide of iron.

N. Alloys.—The most important are those with tin and zinc; when clean iron is plunged into melted tin or zinc with proper precaution, it becomes alloyed on the surface, and hence protected with a film of either of those metals: tin-plate and tinned-iron, and zinced or galvanized iron, is iron thus coated. Iron forms an alloy with nickel, which is interesting from the fact of its nearly always existing in meteoric iron.

O. Tests.—The protosalts are rendered white by ferrocyanide of potassium, and deep blue by ferridcyanide of potassium; the persalts are precipitated blue by ferrocyanide of potassium, not affected by the ferridcyanide, and rendered a deep red with sulphocyanide of potassium. The salts of both oxides form a deep black with tincture of galls; the delicacy of this test is still further increased by the addition of a few drops of ammonia.

20. MANGANESE. Symb. Mn. Eq. 28. (27.67.)

A. Exists native as an oxide. Obtained with great difficulty by intensely igniting its oxide with charcoal; a gray, brittle, somewhat crystalline metal; sp. gr. about 7; very oxidizable, best preserved in naphtha.

B. Manganese and oxygen. There are seven oxides. A protoxide, $Mn\,O$; obtained by passing hydrogen over the ignited peroxide; a greenish-gray powder, very oxidizable; heated in the air it burns like tinder, sometimes pyrophoric; is soluble in acids, and salifiable; its salts, decomposed by alkalies, yield a white hydrated protoxide, which speedily oxidizes and becomes brown.—Red oxide, or Hausmanite, $Mn_3\,O_4$, occurs native.—The sesqui or deutoxide, $Mn_2\,O_3$, is obtained by heating the protoxide in the air, when it acquires oxygen, or by heating the peroxide till it ceases to lose oxygen, or by igniting the protonitrate; it is deep brown, and is salifiable; its hydrate is formed by the action of air on the hydrated protoxide; it tinges glass violet, and appears to be the coloring matter of the Amethyst; it occurs native, crystallized as Braunite, and hydrated as Manganite.—The mineral called Varvicite is a hydrated oxide of manganese, $Mn_4\,O_7 + H\,O$.—The binoxide or peroxide, $Mn\,O_2$, occurs abundantly native, both crystallized and massive; it is the chief ore of the metal; a dark, iron-gray solid; sp. gr. 4.85; it is not salifiable; heated, it evolves 9 per cent. of oxygen, 2 $Mn\ O_2$ being resolved into $Mn_2\ O_3$ and O. 1 lb. consequently yields about 2200 cubic inches, or about a cubic foot and a quarter of oxygen. It is largely used in the manufacture of chlorine for bleaching; bleaching powder, &c. A hydrated peroxide is obtained by the action of bleaching powder on protochloride of manganese; it exists with various proportions of water; with 2 H O it occurs native as Wad.—Manganic acid, $Mn\ O_3$, is formed when the peroxide is heated with nitrate of potassa; manganate of potassa is thus produced, from which the salts of the acid may be prepared; it has not been isolated itself.—Permanganic acid, $Mn_2\ O_7$. Formed when the peroxide is heated with nitrate of baryta; the salt thus formed, decomposed by sulphuric acid, yields a deep red solution of permanganic acid; it may be obtained in carmine red crystals; its salts are mostly red or purple; it is very prone to decompose. Peroxide of manganese communicates a rich purple color to glass, and is employed by glass-makers as an oxidizing agent, and also to counteract the green tint of glass made from common materials contaminated with iron; this purple color, if too deep, may be destroyed by charcoal; it is apt to deepen when the glass is exposed to sunlight.

C. Protoxide of manganese and acids.

i. Protoxide of manganese and nitric acid. A protonitrate, Mn O, $N\ O_5$, is formed by digesting the peroxide in nitric acid with gum or sugar; it forms deliquescent crystals containing 7 H O.

ii. Protoxide of manganese and sulphuric acid.—Protosulphate of manganese, Mn O, $S\ O_3$. Formed direct, or by heating a mixture of the protoxide with sulphuric acid; manufactured by igniting the peroxide with $\frac{1}{10}$th of pounded coal, and dissolving the crude protoxide thus formed in dilute sulphuric acid; may be crystallized with H O—

$4\,HO$—$5\,HO$, or $7\,HO$; a pinkish white, soluble salt, not decomposed by a red heat. It is much used by calico printers as a source of manganese brown; protoxide of manganese, thrown down from it in the fibre of the calico, speedily oxidizes, and forms a rich, permanent, brown dye. Sulphate of manganese combines with other sulphates to form double salts.

iii. Protoxide of manganese and carbonic acid.—Protocarbonate of manganese, $Mn\,O, C\,O_2$. Formed as a pinkish white hydrate, by the addition of an alkaline carbonate to the sulphate; heated, it loses HO, and becomes brown and anhydrous; it occurs native, in pearly, translucent crystals, as spathose manganese.

D. Manganese and chlorine.—Chlorides of manganese. A protochloride, $Mn\,Cl$, is obtained by evaporating the muriatic solution of the carbonate out of contact of air, and heating it to redness in a current of dry, hydrochloric, acid gas; a pinkish white, fusible solid; easily soluble in water, and crystallizing with $4\,H\,O$, or with 2 equivalents of alcohol, which replaces the water of crystallization, forming an alcoate.—A sesquichloride, $Mn_2\,Cl_3$, is formed, when the sesquioxide is dissolved in muriatic acid.—A perchloride, $Mn_2\,Cl_7$, analogous to the permanganic acid; obtained by the action of fused chloride of sodium on a sulphuric solution of permanganic acid; it is given off as a green vapor, condensable at 0° to a deep, olive-colored liquid; it is decomposed by water.

E. Manganese and sulphur.—Sulphuret of manganese, $Mn\,S$. Formed when the protosulphate is ignited with charcoal; a gray, metallic-looking solid; it occurs native, rare; when hydrosulphuret of ammonia is added to a solution of a protosalt of manganese, a pinkish white, hydrated sulphuret is precipitated.

F. Manganese and carbon.—Carburet of manganese. Exists in cast-iron, frequently; occasionally separated in the refining, &c., constituting Kish; supposed to improve steel, when present in small quantity.

G. Alloys.—None of the alloys of manganese are of practical importance, excepting the influence already adverted to, which it appears to produce on iron.

H. Tests.—The salts of manganese are precipitated white by alkalies, the precipitate rapidly becoming brown; they are precipitated white by ferrocyanide of potassium, reddish brown, by hydrosulphuret of ammonia; the deep, though fugitive color produced by oxidation due to the formation of permanganic acid, and the formation of a purple glass when heated with vitreous matters, are characteristic; it is not precipitated by any metal.

21. Cerium. Symb. Ce. Eq. 46.

A. Exists native as an oxide in Cerite, Orthite, and other scarce minerals; very rare. Obtained by heating its chloride with potassium; a gray metal, which decomposes water, and oxidizes rapidly.

B. Cerium and oxygen. There appear to be three oxides; a protoxide, Ce O—a yellow oxide, Ce_9 O_{13}—and a peroxide or sesquioxide, Ce_2 O_3; of these, the proto and sesquioxides are salifiable; the protophosphate and protocarbonate have been found native.

C. Cerium and fluorine.—Fluoride of Cerium. A yellow solid, insoluble in water; it occurs native as Yttrocerite, very rare.

22. Lantanum. Symb. La. Eq. 44.

A. Exists as an oxide associated with cerium; obtained by heating its chloride with sodium; a dark, lead-colored, oxidizable metal (very little examined).

23. Didymium. Symb. D. Eq.

A. Exists as an oxide associated with cerium and lantanum, very rare, (very little examined.)

Section V.—Geology and Geography of Metals.

1. Distribution of Metals in the Strata of the Globe.

   A. Association of metals.

   B. Mineral veins, or lodes.—Indications.—Direction of lodes.—Faults.

   C. General economy of mining operations.

      i. Rock mining;—shaft sinking;—drainage;—ventilation, and lighting of mines;—blasting;—separation of ore from the gangue, or matrix.

      ii. Surface works;—quarrying;—stream works;—sand, gravel, and ore washing;—tin and gold works.

2. Distribution of Metals in different Countries.

   A. Association of ores and fuel;—machinery;—labour;—cost of transport;—economy of fuel.

   B. Mutual dependence of manufactures;—relations with commerce.

   C. Employment of refuse matters.

INDEX.

THE END.